POWER LINES

POWER LINES

GIANT HYDROELECTRIC POWER IN THE PACIFIC NORTHWEST, AN ERA AND A CAREER

Russell McCormmach

Third Edition, Revised

Palimpsest Books
Eugene, Oregon

Palimpsest Books
www.palimpsestbooks.@hotmail.com

First Edition 2009
Second Edition, Revised 2011
Third Edition, Revised 2012

2011935935

ISBN: 978-0-9824110-2-5

For Patrick Henry
Because it was he, because it was I

Electric power is like love – nobody ever gets quite enough.
– newsletter for a Northwest rural electric cooperative after World War II

That big Grand Coulee n' Bonneville Dam'll
Build a thousand factories f'r Uncle Sam

....

What I think is the whole world oughta be run by
E-electricity.
– Song by Woody Guthrie, Talking Columbia

Soul of man
How like the water!
– Song by Schubert, set to a poem by Goethe, Song of the Spirits over the Waters

TABLE OF CONTENTS

PREFACE

The Columbia River and its tributaries have the greatest hydroelectric power potential of any river system in America. Much that is distinctive of the region follows from that fact. The Pacific Northwest has the nation's largest hydroelectric network, supplying two thirds of its electric power, by which measure no other region in the country comes close. The systematic development began with a survey of the Columbia watershed by the US Army Corps of Engineers in the 1920s, which recommended construction of a system of multipurpose dams of unprecedented size, difficulty, and expense. During the Great Depression as part of President Roosevelt's efforts to remove men from the relief rolls, the federal government built two dams on the Columbia River following the Corps' master plan. Through the post-World War II boom years, the government built more dams, while distributing giant blocks of power to the power hungry region whose economy until then had been mainly extractive. The book tells how this vast technological undertaking came about and how it was sustained over decades. With a focus on the Pacific Northwest, it treats of federal power in the dam-building era in America.

The legacy of that era is conflicted. That is seen in recent books about the Columbia River, most of them written from an environmental point of view, critical of massive technological incursions into nature. The hydroelectric development of the Columbia Basin is incontrovertibly both one of the great technological achievements of the twentieth century and a massive incursion. In keeping, this book discusses the development both from the side of engineering, the source of problems in the dam-building era, and from the side of nature, the source of problems in the era that succeeded it, the environmental. The Columbia Basin is home to the country's greatest salmon runs, and dams have severely disrupted them, even as nowhere else has so much work gone into preventing and mitigating the damage. Problems of power and of the environment are of great urgency in today's world, and measures that address them are often contradictory. This book discusses the measures and the place of hydroelectricity among the renewable sources of energy.

As a complement to the historical account, the book includes a portrait of a hydraulic engineer, Mac, whose career coincided with the dam-building era, and who in retirement lived well into the environmental era. A mid-level federal civil servant, he was one of hundreds of engineers like him who designed and built the big dams and the long-distance lines that deliver electric power to the region and beyond.

This is a small book on a big subject. In my belief that power is of concern to everyone now, I have written the book for the general reader rather than the specialist. The first half of the book gives a history of the development of hydroelectric power in the Pacific Northwest. The second half takes up the engineering side of the development; it necessarily contains technical discussions, but readers are not assumed to have a technical background.

I gave an early version of this book as a public talk at Whitman College in 1990, and in 2008 the College asked me to repeat it. For the purpose, I updated the talk, and then I expanded it. As for the timing, I could not have chosen better. The incoming president Barack Obama pledged to make energy a top priority of his administration. Seventy-five years ago the incoming president Franklin Roosevelt made a pledge about energy, and he kept it, as readers will learn. The parallel points up an enduring cardinal fact about a modern economy, the indispensability of plentiful power.

For their helpful comments on this project, I thank Ron Edwards, H. L. Leverton, and Margaret Stewart.

Preface to the Third Edition

This small book on the history of hydroelectric power attracted more interest than I had expected, the result, I take it, of a widening interest in the problems of energy in our society.

The third edition is a thoroughgoing revision. Discussions are brought up to date in light of recent developments, and throughout there are extensive improvements in the content.

I am indebted to Marvin Brammer, John McKern, and Paul C. Pitzer for their generous help. They are not, of course, responsible for any deficiencies of the book or any opinions expressed in it.

PART 1: AN ERA AND A CAREER

SUBJECT AND APPROACH

This book is about the development of a natural resource, rivers, with an emphasis on the hydroelectric power they deliver. Hydroelectricity was one of several major purposes of the Columbia Basin development, the others being navigation, flood control, and irrigation. Typical dams in the basin served all of these purposes, wasting little of value of moving water in meeting human wants. Conceived in the conservation era, Columbia "multipurpose" dams extracted benefits from the rivers that made the region productive in ways that preserved healthy natural resources, and that restored them where they had been squandered.

By singling out one of the purposes, hydroelectricity, I might seem to miss the genius of what was carried out in the Columbia Basin. I give my reasons for narrowing the subject, and here I must get ahead of the story. First, of all the major uses of dams, hydroelectric power was seen as holding unique promise in the Pacific Northwest. The development of the Columbia Basin proceeded from a Corps of Engineers finding that the hydroelectric potential of the Columbia River and its tributaries exceeded that of any other river system in the country. Second, hydropower stood out because of the demand for it. Whereas the demand for navigation, flood control, and irrigation was reasonably settled, the demand for power grew enormously through the period of development. Third, electric power was especially responsive to changing concerns of the region and of the nation. Fourth, hydroelectricity was a strong selling point in obtaining Congressional approval of dams. It was the cash cow of the development of the Columbia Basin, generating revenue that could enable dams to pay for themselves, and that could subsidize other benefits of dams. Fifth, hydroelectricity transformed the region, expanding electricity beyond the cities, and building an industrial base, and it rewarded the nation by strengthening its security. Sixth, Mac, our guide, began his career in a new federal agency created to market hydropower from Columbia dams, and he subsequently joined a new section in the Corps of Engineers, "Hydraulics and Power." Seventh, energy is a major concern most everywhere today. Eighth, and most important, I have chosen to concentrate on what I think is most distinctive of the Columbia Basin development, and most arresting. Federal hydropower in the US, which began as a poor cousin to ancient uses of dams, irrigation and navi-

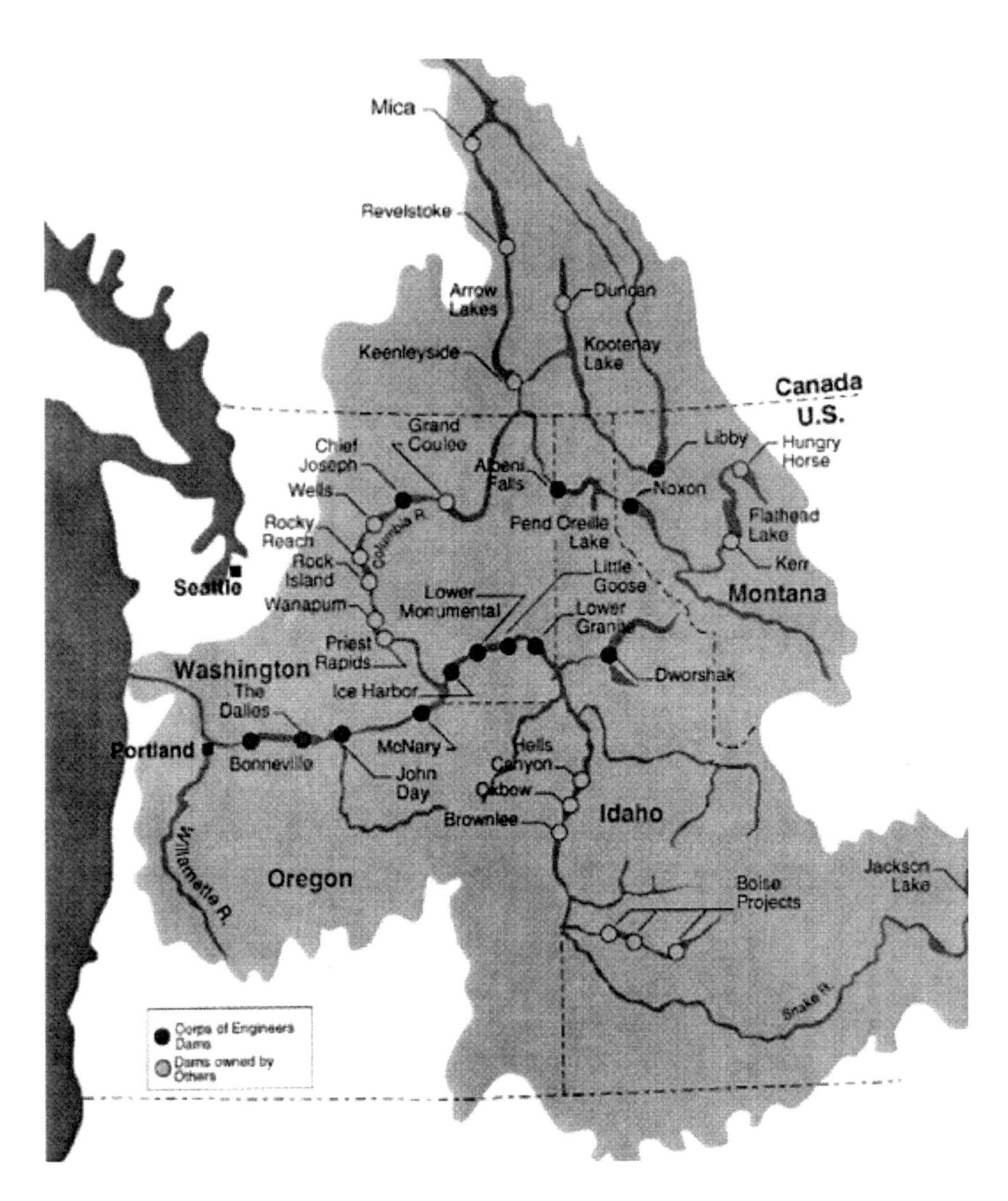

1. Major Dams in the Columbia Basin, Map. US Army Corps of Engineers.

Project	Operator	Location	Year Completed	Type of Project	Authorized Purposes
Libby	Corps	Kootenai near Libby, MT	1973	Storage	Flood Control, Power
Hungry Horse	Reclamation	S. Fork of the Flathead, near Hungry Horse, MT	1953	Storage	Flood Control, Power, Irrigation
Albeni Falls	Corps	Pend Oreille, near Newport, WA	1955	Storage	Flood Control, Power, Navigation
Grand Coulee	Reclamation	Columbia, at Grand Coulee, WA	1942	Storage	Flood Control, Power, Irrigation
Chief Joseph	Corps	Mid-Columbia, near Bridgeport, WA	1961	Run-of-River	Power
Dworshak	Corps	N. Fork of the Clearwater, near Orofino, ID	1973	Storage	Flood Control, Power, Navigation
Lower Granite	Corps	Lower Snake, near Almota, WA	1975	Run-of-River	Power, Navigation
Little Goose	Corps	Lower Snake, near Starbuck, WA	1970	Run-of-River	Power, Navigation
Lower Monumental	Corps	Lower Snake, near Kahlotus, WA	1970	Run-of-River	Power, Navigation
Ice Harbor	Corps	Lower Snake, near Pasco, WA	1962	Run-of-River	Power, Navigation
McNary	Corps	Lower Columbia, near Umatilla, OR	1957	Run-of-River	Power, Navigation
John Day	Corps	Lower Columbia, near Rufus, OR	1971	Run-of-River	Flood Control, Power, Navigation
The Dalles	Corps	Lower Columbia, at The Dalles, OR	1960	Run-of-River	Power, Navigation
Bonneville	Corps	Lower Columbia, at Bonneville, OR	1938	Run-of-River	Power, Navigation

Major Federal Dams in the Columbia Basin. In places, the years of completion of dams in this reproduced table are at variance with years stated in the text, which are taken from better sources. For the effect of the dams on fish runs, the years in which the reservoirs behind the dams were filled are more meaningful. These are: Bonneville 1938, The Dalles 1958, John Day 1968, McNary 1953, Ices Harbor 1961, Lower Monumental 1969, Little Goose 1970, and Lower Granite 1975.

2. Columbia River Rapids. To the west, Mt. Hood is visible. Albert L. McCormmach.

gation, over time became master of the house. By selecting this rags-to-riches story, I do not mean to diminish the other uses of dams, which are also significant for the economy of the region, and in any case I discuss them too.

The Columbia River is the chief river of interest in this account. It rises in British Columbia near the crest of the Rocky Mountains and flows first north and then south through east-central Washington. Just below the juncture with its largest tributary, the Snake River, it bends westward, cutting through the Cascade and Coast Ranges, and forming the boundary between Washington and Oregon. In volume of flow, it is the biggest North American river to empty into the Pacific Ocean. For most of its 1,243 miles, it passes through deep valleys and canyons. It used to have a hundred rapids, it was a wild force. The river is different now, still a force, but no longer wild, its rapids tamed, its swirling and rushing replaced by a settled flow, the channeled stream replaced by a series of lakes behind big dams. No great river has been more changed. For this, the federal government bears foremost responsibility.

Why, we may ask, would the government do that to one of its most beautiful rivers? Today we know that big dams can harm the environment and disrupt the natural order. They can cause species to decline. They support industries that can pollute the earth. They often encourage cash crops through irrigation, in return depleting the soil and impoverishing the population. They often drown out floodplains, which otherwise absorb floods, or conversely, they allow floodplains to be settled, increasing the damage of the next big flood. These and the many other unwanted effects of big dams make for a long list. Young persons who have grown up in the environmental era wonder how all this could have happened. They wonder, that is, what those early humans could have been thinking about. This book takes up the question.

I come to the subject as a historian with a historian's way of thinking about the past. I have found that people usually have their minds made up about the

3. Columbia River Rapids. US Army Corps of Engineers.

present but that their minds are open about the past. That itself is a good argument for history. Historians often choose to study certain events from the past because they are of particular concern in the present. That way they might open minds to the present. History has a modest part to play in today's debates.

Discussions of the development of the Columbia River watershed, or "Columbia Basin," dwell on the plight of fish. That selection does faithfully reflect our main concern today. Fish were not absent from the concerns at the time of the development but they were not a main concern. From our vantage, we might say, the shame. Had it been otherwise, our problems today would no doubt be different. In the time of the development, the main concerns were economic and engineering, no surprise there. Given our concerns today, it is no surprise either that most discussions of the development pay scant attention to the engineers who did the work. This small book is in part an acknowledgment of the gap.

To avoid the familiar pitfalls of writing about a large development, I give the Columbia Basin a human face. It takes the form of a mid-level federal civil servant, a draftsman and hydraulic-design engineer, known to his colleagues as Mac and to me as father, Albert L. McCormmach. Mac's career coincides with the period of the major development of the Columbia Basin.

Mac was employed by the civilian branch of the US Army Corps of Engineers, which built most of the big dams in the region. When he retired from government service, he left all of his drawings, computations, and memoranda at the office, taking with him only a loose-leaf album of photographs, mainly photographs of dams. These photographs gave me the idea for this book. I include a generous selection from them; I regard them as equally important as the text.

They have to do with technical problems that came up in the construction and operation of dams, though the care Mac took in mounting and indexing them may point to a further reason why he kept them. The engineering work on the Columbia is significant, and the photographs are a part of its historical record. When Mac went on field trips, he took photographs with his own camera and also with cameras on loan from the Corps. The photographs he took with the Corps' cameras he gave to the Corps, which developed the roll of negatives and then made a set of prints for him and a set for itself. The Corps kept those negatives. Mac kept the negatives of the photographs he took with his own camera. The album also contains photographs of dams taken by professional photographers of the Corps, who recorded every stage of construction. All of Mac's photographs for which I do not have negatives I credit to the Corps, though some of them may be photographs Mac took using a Corps camera. With exceptions, the photographs reproduced here were taken by Mac and by Corps photographers.

Certain of the photographs have intrinsic interest, such as those showing the rhythmic pattern of spillway bays, but that is incidental to their purpose. Mac, who had been a serious amateur photographer in his younger years, thought that their photographic interest was under-appreciated. Photographs were not prominently displayed in the building where he worked, but filed away. The occupants of the building were engineers, as Mac explained it, who care about how things work, not how they look.

The written account accompanying Mac's photographs is based upon Mac's recollections, on letters he wrote at the time, and on two tape recordings, one undertaken at my prompting, the other giving his impressions and thoughts on a solo drive down the Columbia River late in life. The account also draws heavily on publications by historians and other writers dealing with the Columbia Basin development.

I consider this book a historical sketch and memoir with historical photographs. Although it is documented, it is not a work of original scholarship. Other than for material on Mac's career, the information it contains can be found in other published sources. For readers who want more than an outline of the subject, there are detailed studies, several of which I list at the end. The novelty of the present book is its point of view, that of a time of power development in the Pacific Northwest together with the career of an ordinary engineer who had a small but necessary part in it.

Because the scale of the development is important, I give numbers throughout the book, possibly more than some readers want, I realize. More of a block to some readers may be frequent references to government bureaus and acts of Congress. This cannot be helped, since the development of the Columbia Basin was largely a federal undertaking. This book contains descriptions of problems that came up in the design of dams, but it is not a history of engineering. The technical discussion is elementary and should not give readers any difficulty.

I grew up with dams. In addition to sharing the conflicted feelings my readers may have about issues surrounding dams, I come to this subject with the additional conflicted feelings that one has about another member of the family, in my case dams. I do not approach the subject indifferently but I trust fairly. The early part of this book may come across as a highly unfashionable valentine to an era of power, but if anyone finishes the book without feeling the poignancy of Mac's career, I will have failed as a writer.

The book is divided into four parts. The first part is an account of the era together with Mac's career. The second part is offhand observations that Mac recorded on a trip down the Columbia River after he retired. The third part is technical, an explanation of dam construction and a description of Mac's work. The last part is about individual dams, with many photographs.

POWER

Electricity is a late comer in the history of water power. The use of the power of moving water goes back to antiquity. The Greeks and Romans, we know, built water-powered mills to grind grain. Water power was still the main source power for machinery through the Industrial Revolution of the eighteenth century, though a new, revolutionary kind of power made its appearance then, steam, fueled by coal. The nineteenth century added another new kind of power, one more revolutionary in the long run, electricity. By the time the Pacific Northwest began to develop an industrial base, the age of electric power was well underway. An era of construction of big hydroelectric dams in the Columbia Basin began in the 1930s and ended effectively in the 1970s, and definitively in the 1980s. Big dams brought wealth to the region, and they also brought daunting problems, which still haunt us.

Recall that of the multiple uses of big dams, we are concerned mainly with the generation of power. We should be clear about the meaning of the terms we use. The first meaning of "power" is the capacity to cause an effect. A narrower meaning is a physical agency for supplying energy; or, to be more technical, it is the time rate at which energy is transferred or emitted. The primary physical agency we are concerned with is moving water. The secondary physical agency is its electrical equivalent. To the dams of the Columbia Basin, which bring about the conversion between the two forms of energy, we give the hybrid name "hydroelectric dams." (In this book, I usually speak of "power," but sometimes I speak of "energy." Others do this too. The World Power Conference in 1968 changed its name to World Energy Conference, a more accurate reflection of the organization's interest in the entire spectrum of energy.) "Power lines," the title I give this book, in its straightforward meaning, refers to the network of high-voltage transmission lines that convey the electric power generated at the dams of the Columbia Basin to the four corners of the Pacific Northwest and beyond. It has a borrowed meaning too, which comes up later.

A second, common meaning of "power" is the ability to affect the environment, including human affairs. What evolution is to biology, energy is to the physical sciences, the great generalizing concept, and it has been carried over to the human sciences with similar ambitions for it. In anthropology, there is an energy theory of cultural evolution. In psychology a well-known theory of personality emphasizes the dynamics of power. The philosopher Bertrand Russell wrote a book *Power* "to prove that the fundamental concept of social sciences is power, in the same sense in which energy is the fundamental concept in physics." (He did not succeed, but like everything he wrote, his book sparkles with wit, and its blunt truths are still capable of startling.) The subject of the present book, the development of hydroelectric power in the Columbia River Basin, brings together the two halves of the analogy, physical power and human power. (Of course, humans have physical power too.)

Wherever we look in history, we usually see two elements present, power and change, which are entangled. Change can be gradual enough to be imperceptible and it can be rapid enough to overturn society, and it can be large enough to mark a new era in history; when that happens we find new forms of power in society. Society rarely if ever rejects new forms of power, whether human or physical. Rather it welcomes them, and usually without many scruples. The civilized arts have arisen hand in hand with them. They impart to history its restless change and fascination. They liberate, and they disrupt at the same time; they invariably arrive with unintended benefits and costs. The balance sheet for the hydroelectric development of the Northwest has still not been totaled. Just as new powers carry considerable risk, they make their appearance in times of considerable risk. The years of the development of hydroelectric power were also the years of the Great Depression, World War II, and the tensions of the Cold War. They were also years of great cultural, economic, and political change in the nation and in the world, much of it humanly and materially beneficial.

With every new physical power there come questions, by and large the same questions. Who controls the power? What is it for? How much does it cost? With the arrival of the new physical power hydroelectricity in the Pacific Northwest, the answers to these questions came after struggles and compromises among human powers, as always.

"Power lines" in its borrowed, human meaning refers here to the clash of interests of groups affected by Northwest dams: barge owners, irrigation farmers, Indians, aluminum manufacturers, hunters and fishers, government agencies, environmentalists, politicians, power managers, and still others. Once in place, built of steel and concrete, dams and towers that carry their power convey authority and permanence, as if placed there by divine edict. They give no hint of the human powers that actually put them up . . . or that can take them down.

DAMS

Hydroelectricity largely comes from dams, especially big dams. The early truly big dams in the US such as Wilson Dam, Hoover Dam, and Shasta Dam, all of the dams on the Columbia and Snake Rivers, and most of the other big federal dams were justified by revenue from hydroelectricity. Today most hydroelectricity comes from big dams like these; the Pacific Northwest has 160 hydroelectric projects, but fifty percent of the hydroelectricity in Oregon, Washington, and Idaho comes from only the six biggest dams there.

In this section, I use the expression "big dams" loosely, as I do throughout, without defining it, since the measure of bigness changes with the times. For this book, the classification of dams is not an issue, and I see no reason to adhere to the international standard terminology, but readers should know that there is one: "large dam" is a dam standing over fifteen meters, about forty-five feet, and "major dam" is a dam standing over 150 meters, about 450 feet. All of the dams discussed in this book are large or major dams. They are a significant subset of the roughly 50,000 large and major dams in the world today.

With our choice of subject, structures built to control major rivers, we are definitely bucking a trend. With an over-crowded planet, and with an overburdened environment, we are urged these days to think small, and about nothing more so than the scale of our technologies. Giant technology is decidedly out of fashion, "nano"-technology is in, and the technology of everyday things is somewhere in between and down-sizing. But there are signs that the trend may reverse. There is a field geo-engineering that considers solutions to urgent problems such as global warming on a scale that dwarfs our biggest dams.

Dams on rivers can be natural, caused by lava flows and landslides, but we are concerned only with artificial dams. Artificial dams first appeared around

4. Hoover Dam. Originally named Boulder Dam, this early multipurpose, 726-foot-high dam on the Colorado River was built by the Bureau of Reclamation between 1931 and 1935. The period is reflected in the art deco facade. Hoover Dam holds the largest reservoir of any dam in the US, and until Grand Coulee Dam on the Columbia River, it had the greatest hydroelectric capacity of any dam in the world. Pamela McCreight.

6,000 years ago in the Middle East and Mesopotamia, and so did civilization, and the two have evolved together ever since. Early dams were used to control water levels. Down the centuries their functions have come to include recreation, city water, field irrigation, water for industry including storage of industrial effluents, boat passage, flood management, and power generation. The dams discussed in this book are, as mentioned, "multipurpose."

5. Grand Coulee Dam. Located on the Columbia River in north-central Washington, this multipurpose, 550-foot-high dam was built by the US Bureau of Reclamation between 1933 and 1942. It has the largest hydroelectric power capacity and holds the sixth largest reservoir of any dam in the US. This photograph shows the dam after it was expanded to accommodate a third powerhouse, on the left. US Bureau of Reclamation.

There were a few early big dams. In the fourteenth century, the Mongols built an arch dam over 180 feet high in a narrow canyon in Kurit, Iran. With its height increased by twelve feet in 1850, the Kurit Dam was the world's highest until early in the twentieth century. Built for flood control in 1866, Furens Dam in France was nearly as high, measuring 184 feet on its downstream face, still a giant for its times. By the turn of the twentieth century, there were lots of dams in the world, but very few stood as high as fifty feet. In the early years of the century, improved methods of analysis and improved materials and construction techniques made it possible to build much higher dams. American dam-builders took the lead. Foremost among them, the Bureau of Reclamation built a series of high irrigation dams in the West: 214-foot Pathfinder Dam on the North Platte River in Wyoming in 1909; 325-foot Shoshone Dam on the Shoshone River in Wyoming in 1910, briefly the highest dam in the world; 280-foot Theodore Roosevelt Dam on the Salt River in Arizona in 1911, the biggest masonry dam; then 349-foot Arrowrock Dam on the Boise River in 1915. The Bureau picked up the pace in the 1930s with 417-foot Owyhee Dam on the river of that name in Oregon, in 1933, then the world's highest dam; towering 726-foot Hoover Dam on the Colorado River in 1936, the record dam; and 602-foot Shasta Dam on the Sacramento River, begun in 1938. By the time of Shasta Dam, the Corps of Engineers had become a partner and competitor in the construction of big dams. In the decade of the 1930s, the four biggest concrete dams in the world were under construction at the same time in the West: in addition to the Bureau's Hoover and Shasta Dams, they were the Corps' Bonneville and Grand Coulee Dams on the Columbia River. Hoover Dam came first in time and remained the highest dam; in volume of concrete Shasta Dam was half again as big as Hoover Dam; and Grand Coulee was bigger than both dams together. Under construction at the same time in Montana was the Corps' earthfill Fort Peck Dam, which except for the Great Wall of China was the largest man-made structure in the world; *Life Magazine* featured Fort Peck Dam on the cover of its first issue in 1936. What skyscrapers had been to Americans in the 1920s – the Empire State Building, the Chrysler Building – big dams were in the 1930s. The symbolism at the time was significant. Many more big dams were built in America in the 1950s and 60s, if on a less monumental scale than Grand Coulee, and many more and bigger dams were built in other countries. In mid-century, when the Army Corps of Engineers started up a new district in Walla Walla, Mac's employer, to design and operate new big dams in the Columbia Basin, there were over 5,000 big dams in the world.

The strong economic growth in America and other parts of the world after World War II was accompanied by an escalating demand for power and an extraordinary spate of big-dam building. In the decade of the 1970s, the peak, over 5,000 big dams were built, or an average of 500 dams a year. Twenty years later, though the pace had slackened, still around 200 big dams a year on the average were started. Today in the United States and Europe, big dams are rarely built anymore. Rather there is the converse trend of decommissioning dams, occasionally big dams, for economic or environmental reasons, but elsewhere in the world, the building era has far from run its course, and the problems with big dams have a renewed life.

Take China. Big dams were promoted by Chairman Mao, who built the 350-foot-high Sanmenxia Dam on the Yellow River as a monument to man's mastery of nature. The river, true to its name, in no time filled the reservoir with yellow silt, requiring another big dam to correct the mistake. Sanmenxia Dam has caused as many floods as it has prevented, and has ruined as many lives as it has improved. This mistake has not slowed China. The Yellow River has twenty big dams today, and eighteen more are scheduled to be built by the year 2030. It is the same on the Yangtze River. Today China has half of the world's big dams and leads the world in building big hydroelectric dams, though it also makes wide use of small-scale hydroelectric power generation, which looks to be the wave of the future everywhere. While China is building dams, it is also building coal-burning power plants at an unprecedented rate. The

6. Three Gorges Dam. In a murky sky. Located on the Yangtze River in China, this 331-foot-high dam was begun in 1994 and completed, as of the original plan, in 2008. Because additional generators are yet to be installed, it is not expected to be fully operational until 2011. Its total hydroelectric generating capacity at that time will be 22,500,000 kilowatts, roughly twice that of its nearest rival, Itaipú Dam. Christoph Filnkössl.

environmental problems these alternatives to hydroelectric power pose for China and the rest of the world are well-known too. China is committed to growth, and it needs power to lift the mass of its people out of poverty. Its leaders know that the survival of the regime probably depends on this.

What happened in America first, and what is happening on the Yellow and Yangtze Rivers, is happening on many other major rivers round the world. Grand Coulee Dam, which at one time had the greatest hydroelectric capacity in the world, now lags in fifth place behind Three Gorges Dam on the Yangtze River, Itaipú Dam on the Paraná River in Brazil, Guri Dam on the Caroni River in Venezuela, and Tucuruí Dam on the Tocantins River in Brazil. Dams as high as a thousand feet have been built outside America. It is well enough for us to say that building big dams is obsolete, that it is too late in the environmental history of the earth to burden it with more of them, but we have our big dams, and for people who live in other parts of the world the power is there for the taking. Two thirds of the world's big dams are in developing countries. In time, Beijing and other capitals will lose interest in building big dams too, and for the same reasons, but not before big dams have been built on most of the promising sites.

7. Itaipú Dam. This dam was built in 1971-84 as a bi-national project. Located on the Paraná River between Brazil and Paraguay, it supplies Brazil with twenty percent of its energy and Paraguay with over ninety percent. Its hydroelectric power capacity is second only to that of the Three Gorges Dam in China. Tom Oates.

From the start, the big-dam development of the Columbia Basin had a world setting and world support. The first Congress of the International Commission on Large Dams met in Stockholm in 1933, the year the construction of federal dams in the Columbia River began. The US was one of the twenty-one members, and by a relevant index the most prominent, with the biggest dam in the world, Boulder/Hoover Dam, well underway and competitors in the planning. On the Columbia River and in the world at large, big dams and electric power went hand in hand. The International Commission on Large Dams was a spinoff of the World Power Conference, and at its constitutive meeting it was endorsed by the International Congress of Electricity Producers and Distributors. The International Commission is still going strong, with eighty-eight members, holding congresses every three years in different countries. It believes in big dams.

We encounter critical judgments on big dams all the time; it is instructive to step onto the turf of supporters of big dams. The International Commission on Large Dams, for example, takes a long historical perspective, pointing out that for thousands of years dams and reservoirs have sustained human life, and that today as ever human life is threatened by an imbalance of water, food, and energy. Large reservoirs hold about forty percent of the annual global runoff, and thirty to forty percent of global irrigated land relies on dams. There are already a half million reservoirs covering 2 1/2 acres or more, and given the growing demand, especially in the developing countries, the world needs still more, the Commission says. To date, only a modest fraction of the world's renewable water resources is regulated by dams and reservoirs.

BEGINNINGS, EARLY HYDROELECTRIC POWER IN AMERICA

We who have grown up with electric power have little feeling of what life was like without it, when the sources of power were human and animal muscle, fuels, and moving water and air. Electric power is instantaneously available, precisely controllable, dependable, convenient, quiet, effortless, and big, and it is convertible into other forms of power, chemical, thermal, and mechanical. For convenience and versatility, there is no other power to compare with it. The changes it brought about in everyday life in the first three decades of the twentieth century have been called a "revolution," and with good cause. They included the replacement of fuel lamps by electric lights on the street, of steam power by electric power in the factory, and of hand labor by a range of appliances in the home such as the washing machine and the electric iron. They included the radio for entertainment and information. They included electric lights, which were brighter and far more convenient than previous lights; electric motors were more powerful than steam engines, and electric wires dispensed with the cumbersome shafts and belts of mechanical transmission; and with the introduction of alternating current and electric motors, most of the standard home appliances as we know them came on the market in rapid succession.

Unlike winds and rivers, electricity is not given to us freely by nature. It takes power to produce it, and water is one of the means. Hydropower appeared very early in the history of electric power in America: it was first used in industrial lighting in 1880 at Grand Rapids and in city street lighting at Niagara Falls the following year. The first hydroelectric power in the Northwest was carried by a fourteen-mile transmission line from Willamette Falls to Portland in 1889, again for city street lighting. Early hydroelectric plants took advantage of natural waterfalls to drive the water turbines that ran the electric generators. But waterfalls were few and far between, and they were unsteady. Dams were the future of hydroelectricity, even where there was a natural falls. Dams provided a more reliable flow through the turbines than nature.

The early demand for electricity was met largely by electricity generated at factories and utilities using coal, but as hydroelectric plants multiplied they complemented coal-fired plants. Utilities, which at first were "light" companies, became "power and light" companies as they increasingly powered industry as well as residential and street lighting, and their names sometimes reflected the change. For reasons of efficiency and continuity of service, utilities showed a tendency to consolidate from very early, held back only by the shortage of power-line hookups. Pacific Power and Light Company in Portland was formed in 1910 through the consolidation of troubled utility companies in several towns in the Pacific Northwest. Seattle City Light was formed the same year as a municipal utility, in charge of the first municipally owned hydroelectric dam. (Its superintendent beginning next year was J. D. Ross, who built several more hydroelectric dams for the utility, and who would go on to be the first administrator of the Bonneville Power Administration, the agency created to distribute power from the first federal dam on the Columbia River.) In 1912, Seattle Electric became Puget Sound Traction, Light & Power Company ("Traction" later being dropped, and then "Light," the company then being referred to simply as "Puget Power," going by the most important word). Marketed by rapidly growing numbers of consolidated utilities, hydroelectricity came to supply nearly half of America's power in the early twentieth century. For this reason water power was called "white coal," which together with black coal was responsible for a good share of America's industrial dynamism.

Throughout the country, public power in the beginning was concentrated in cities like Portland and Seattle; private power developed alongside it, at the same time and in competition, generating a controversy that would continue through the period this book treats. Investor-owned utilities would serve far more Americans than public utilities; in this respect, the Pacific Northwest was somewhat of an exception, public utilities being given preference in the marketing of federal power.

During World War I, the federal government gave utilities priority over factories in the generation of power. The transport of coal to factories was costly, and it took a great deal of scarce rolling stock. It was more efficient to deliver coal only to the central generating plants owned by the utilities and to let the utilities supply the factories with electricity over long-distance transmission lines. The government required only that there be a tie-in between existing power lines to allow excess production to be shifted; the first interconnection in the Northwest was made in 1918 between Puget Power and Washington Water Power. Utilities envisioned a large, interconnected system of new and existing power-generating plants they could tap into; they called this "superpower," and the idea took hold There were social reformers at the time who had a more ambitious plan, which one of them called "giant power." Their idea was to have a nationwide interconnection of regional transmission grids, freeing Americans everywhere from the drudgery of labor through electric power. It was understood that this plan called for public control over electric power, and this frightened off the utilities. It was left to the federal government to take up key features of giant power: public control, central generating facilities, and wide distribution of electric power, especially in rural areas. This happened in the Columbia Basin, and the Columbia giant grew ever bigger. The subtitle of this book is a borrowing from this form of giantism.

The federal government produced some of the early electric power in America. The Bureau of Reclamation was authorized from the start to develop the hydroelectric potential of its irrigation projects. It early on fell into the practice of building hydroelectric plants at its dam sites to power machinery. In 1906 Congress further authorized the Bureau to market the surplus electricity from Theodore Roosevelt Dam on the Salt River in Arizona; in 1909, before the dam was completed, the Bureau began supplying nearby Phoenix with residential and commercial electricity from the dam. Likewise in other parts of the arid West, including places in the Pacific Northwest such as Yakima, Washington and Umatilla, Oregon, the Bureau provided cheap hydroelectricity to farms, industries, and cities; this by-product of federal irrigation works was the beginning of the federal production of electric power.

Like the Bureau of Reclamation, the Corps of Engineers arrived at hydroelectric power through the back door. During World War I in response to an urgent need for munitions, President Wilson approved the construction of two nitrate plants and a dam to supply them with electricity at Muscle Shoals on the Tennessee River. The Corps of Engineers was assigned to build this dam, Wilson Dam; when the war ended, the Corps was not yet finished with the construction. The rightful ownership of this wartime facility in peacetime, whether it was to be retained by the government or to be sold to private industry, was passionately debated in Congress. The issue was resolved only in 1933,

8. Theodore Roosevelt Dam. Until 1959 named Salt River Dam, Theodore Roosevelt Dam was built by the Bureau of Reclamation between 1903 and 1911 to control the flow of the Salt River and to irrigate the Arizona desert. As a by-product it developed a modest amount of hydroelectricity as a "paying partner" of a water reclamation. The Salt River Project was the first multipurpose reclamation project, a forerunner of the Bureau's multipurpose Grand Coulee Dam and Columbia Basin Project. At the time it was built, it formed the largest artificial reservoir and at 280 feet it was the highest masonry dam in the world; its height has since been raised to 357 feet. It was the first major Bureau of Reclamation project to be completed, and it marked the beginning of federal hydroelectric power production. McCulloch, Library of Congress.

when Roosevelt absorbed Muscle Shoals together with Wilson Dam into the new Tennessee Valley Authority. In the meantime the Army Corps of Engineers had completed Wilson Dam, then the biggest concrete dam in the America, and one of the biggest producers of hydroelectricity. In partnership with private power companies, the Corps helped build several more dams in the first years after the war. It was as a somewhat experienced hydroelectric power developer that it turned its attention to the Columbia River in the mid 1920s.

The legal framework for the development of hydroelectric power in America was laid early in the century. Hydroelectric projects were to be integrated into comprehensive plans for American waterways, and returns from the sale of hydroelectric power were to be used to pay for the cost of the dams and transmission facilities and to finance navigation and flood control projects connected with the dams. The Army Corps of Engineers, whose responsibil-

9. Wilson Dam. This 137-foot-high, early multipurpose dam on the Tennessee River in Alabama was built by the Army Corps of Engineers in 1918-27. At the start of Roosevelt's New Deal, it was transferred to the Tennessee Valley Authority, becoming one of its eventual nine dams on that river. Tennessee Valley Authority.

ity it was to approve the sites and plans of all dams, was to see to it that the legal requirements on hydroelectric power were met. The Federal Power Commission, a licensing body created by the Federal Water Power Act in 1920, was authorized to issue licenses for hydroelectric projects on public land, and in its first year it ruled that the federal government could build dams on the Colorado and Columbia Rivers. The river giants Hoover, Bonneville, and Grand Coulee Dams were the first results of this ruling.

During his first term in office, 1933-37, Roosevelt strengthened the Federal Power Commission, and he backed legislation creating the Tennessee Valley Authority, the Bonneville Power Administration, and other power-dealing agencies. The New Deal proved a watershed in the history of electric power in America, the federal government becoming a force in electric policy for the first time. The legal, political, and governmental conditions for the sustained hydroelectric development of the Columbia Basin were in place.

WHY THERE ARE DAMS ON THE COLUMBIA RIVER

We can build hydroelectric dams to work for us, but since they are costly a convincing case has to be made for the work they do. That was the problem with Columbia River dams when they were first proposed. They could produce electric power, but the region lacked the population – it was only around three million then – and the industry to benefit from it. The president of the American Society of Civil Engineers thought that Grand Coulee Dam would be as useless as an Egyptian pyramid.

Columbia River dams made good engineering sense, but not yet convincing economic sense, and economics came before engineering. Actually, the dams occupied a gray area in economic thinking. They were intended to serve a number of uses at once, and the uses depended on contingencies that could not always be assigned dollar values in advance. The dams were a long-term investment, and their uses could change in unforeseen ways over time. A future could be imagined in which the dams would pay off handsomely. Since federal dams had to be approved by Congress and the administration, political considerations in the end decided whether or not a dam would be built, and these considerations were sometimes only loosely tied to economic ones.

The motivation to build the first federal dams on the Columbia River came not from hard economics but short-term, local unemployment relief combined with a range of hopeful desires. While the construction of those dams was underway, world events intervened, after which demand began to catch up with desires. In time, the hydroelectric power produced by the first dams attracted power-users to the region, which stimulated an economic need for more dams like them. The hydroelectric development of the Columbia Basin arose from, and was sustained by, a combination of opportunity and means, and the catalyst was desires. We look first at desires.

1. Desire to build. This desire is as old as civilization, and it is not only for a roof over our heads but it is as well for buildings that make a bold impression on us: temples, pyramids . . . and big dams. Big dams are impressive objects, and what they do, control the flow of mighty rivers, is no less impressive. Visitors on tours of Columbia Basin dams are impressed, and they often find themselves smiling, whatever their private reservations. Looking down from the top of a dam on water cascading over the spillways or watching and feeling the vibrations and hearing the roar of the turbines spinning in the powerhouse are experiences like no other in life. Visitors are filled with wonder, quite helplessly. I have been there with visitors, and I have observed them, and I have had the same feeling.

2. Desire for power. For sustained material progress, long assumed to be an

American birthright, generous quantities of physical power must be available for production and for living. Of the kinds of physical power, the one that Americans came to prefer in the twentieth century is electric power – clean, convenient, and ever so versatile. There grew an resistible desire to enjoy its blessings. Given this and the thinking of the time, and given the technology for generating and distributing electricity, the fate of America's paramount power stream, the Colombia, was all but sealed.

3. Desire for progress. Like the desire for power, the desire for progress touches on a number of others in this list. In the Columbia Basin, it was the desire to put to use the latest technology for the betterment of society; the technology for building very big dams was recent, and some of the moving parts of dams were advanced for their time. It was also the desire to use electricity to bring about progress in general: dams delivered electricity, which was involved in nearly every new product that people counted as material progress.

4. Desire for security. Production goals and power shortages in America during the First World War demonstrated the importance of electric power for national defense. The world once again was becoming a dangerous place when the first federal dams on the Columbia, Bonneville and Grand Coulee, were under construction. This was the eve of World War II, and the Axis countries had three times as much power available for armaments manufactures as the United States. The administrator of the Bonneville Power Administration told Congress that modern war is fought in the factory no less than in the air and the trench, and that electric power supplied by the new dams in the Pacific Northwest would meet any national emergency. In keeping with this thought, the secretary of war approved the installation of additional electric generators at Bonneville Dam and at other dams in the West. The Bureau of Reclamation, the biggest supplier of power to defense industries in the West during the war, quadrupled its hydroelectric output between 1941 and 1944. Grand Coulee and Bonneville Dams and their major transmission lines were finished in time to supply enormous blocks of electricity for essential metallurgical and electrochemical manufactures during the war. They powered plants that produced the aluminum to build 50,000 warplanes, and they powered shipyards that turned out a liberty ship a day. At a time when rail transport was scarce, the lock at Bonneville Dam allowed river transport of munitions, grain, and other wartime products. In his bunker, Hitler despaired of America's "ceaseless production." In Tehran, Stalin, in a rare complement to America, toasted, "To American production, without which this war would have been lost." Persistent doubts about the justification of the first Columbia River dams did not survive the war.

5. Desire for autonomy. Northwesterners tended to think of themselves as

poor cousins of Easterners, or thought that that was how Easterners thought of them. Easterners had exploited them for their resources and controlled them financially long enough. They wanted to be an autonomous agricultural and industrial power similar to, but independent of, that of Easterners. Dams looked to be a route to that end.

6. Desire for earthly paradise. This desire, also ancient, took a characteristic American turn in the Northwest, the family farm. To the pioneers crossing the Inland Empire, the popular booster name for the high plateau in the Pacific Northwest stretching from the Cascade Mountains in the west to the Rocky Mountains in the east, was a hot, very arid wasteland, an obstacle in their path to the promised land, the Willamette Valley. To spread the promised land inland, the Bureau of Reclamation developed a bit of the wasteland in the Yakima Valley in central Washington early in the century. This became its model for the Columbia Valley, with a difference: Yakima water does not have to be lifted far to irrigate the surrounding land; Columbia water, by contrast, has to be lifted out of deep canyons in many places. From an idea floated in a newspaper in 1918, local promoters of irrigation in eastern Washington lobbied for a federal dam on the Columbia. The idea was to use the power of the river to run giant pumps to raise water to the level of the land and to store it in a vast natural channel, Grand Coulee. Without water this land was forbidding, but with water it was a rural paradise. The promoters of Grand Coulee Dam envisioned an irrigated, flood-controlled Garden of Eden for a large, imported population of resourceful Americans, who lived on the land they worked. What came of their paradise is discussed in a later section on Grand Coulee Dam.

7. Desire for river passage. Rivers have been a major artery as far back as records go; and governments invest money and enact laws to facilitate commerce and transport on them. The federal government first became interested in the Columbia River as a navigation stream, dredging and building locks and canals. Later, when it built dams on the river, it built locks at the same time, since together they improve navigation. The government's first dam on the Lower Columbia River is called "Bonneville Lock and Dam," the emphasis given to the lock; but for Grand Coulee and Chief Joseph Dams, all government dams on the Columbia are named the same way, lock before dam. Because of the locks and dams on the Columbia, it is now far cheaper to transport wheat downstream and petroleum upstream by river than by land. The locks and dams have benefited the region in this way, a desire fulfilled.

8. Desire for safety. This desire is rooted in humanity's primitive fear of natural disasters. Big floods account for forty percent of natural disasters in the world; many thousands of lives are lost to floods in a typical year. We build dams as a means of river control, and up to a point they protect us. Reservoirs

created by them hold back water during floods and release it when the danger is past. All federal dams in the Columbia Basin have a measure of flood control. But with big floods, all bets are off. We cannot control them or any other natural disaster for that matter. We know this, but we underplay the dangers. While the sun is out we build our houses on the flood plain. Like us, most of the time dams belong to a world where the sun always shines, where uncontrollable nature is relegated to perpetual night; dams are reassuring.

9. Desire for River Uses. We have come to desire from rivers everything they have to offer. In the Progressive era, there was a movement to introduce scientific efficiency into the management of the water resources of America. It was about conservation, and it found a vigorous champion in Theodore Roosevelt. In a report to Congress on rivers Roosevelt said, "It is poor business to develop a river for navigation in such a way as to prevent its use for power, when by a little foresight it could be made to serve both purposes ... Every stream should be used to the utmost." His distant relative Franklin Roosevelt, while campaigning for the vice presidency in 1920, visited the Columbia River Gorge. "I could not help thinking," he said, "as everyone does, of all that water running unchecked down to the sea." The desire for multipurpose river projects to realize the full benefits of rivers caught on around the time of Roosevelt's visit. The Corps of Engineers' first multipurpose project, Wilson Dam on the Tennessee, which combined hydroelectric power with navigation and flood control, was then under construction. The desire found full expression in the Rivers and Harbors Act of 1925, which authorized the secretary of war to use the Corps of Engineers and the Federal Power Commission to formulate a plan and estimate costs for the improvement of navigation on American rivers with hydroelectric potential, and at the same time to take into account flood control and irrigation. This Act laid the groundwork for federal multipurpose dams on the Columbia River and its tributaries. President Franklin Roosevelt started the development by funding a multipurpose dam at the place he had visited on the Columbia River, the Columbia River Gorge.

10. Desire for economic growth and recovery. Of the various desires behind the construction of dams, one was paramount, a desire for wealth. The development of rivers, as of other natural resources in America, was traditionally tied to economic growth, and private enterprise and governments at every level looked to dams to generate wealth from rivers. On the Colorado River, Hoover Dam, originally named "Boulder Dam," was undertaken in the 1920s to promote economic growth, and plans for building more big dams for the same purpose were under consideration when the Great Depression arrived. The extractive economy of the Northwest was especially hard hit; by 1932, for example, eighty percent of the region's lumber mills had shut down. Power and

light utilities, which had flourished during the inter-war years, retrenched for lack of capital for construction. Roosevelt, during his presidential campaign, as we have seen, promised to develop the Columbia River to put people in the Northwest back to work. Most of the construction crews at Bonneville and Grand Coulee Dams came from Oregon and Washington relief rolls, as intended. Jobs were a short-run justification – when the dams were finished, the jobs were over, and the dams took very few persons to operate – but there was an emergency in the region, and the short-run counted. In any case, the future would take care of itself, Roosevelt was confident of that. Hard times and an adventurous administration started the ball rolling. The era of big-dam building in the Columbia Basin owed to a robust optimism in the depth of the Great Depression.

As river projects, dams on the balance were considered a sound investment. They had low maintenance and operation costs, and their technology had proven reliable over time. Their bulk was thought to replicate natural dams, and their performance to imitate natural processes. They were looked upon as collaborators with nature as much as conquerors of nature. When the development of the Columbia Basin began, dams were widely regarded as productive and protective, as instruments of conservation and restoration. If the idea of friendly dams sounds strange to ears accustomed to environmental objections to these incursions into nature, it is a measure of the distance between their era and ours.

Different parties had different desires for the development of the Columbia Basin, and their desires often conflicted with one another. And there were variants on the desires, and still other desires, but the ones above were more than sufficient. Because of the expense of big federal dams, their justification had to be put in terms designed to get legislators' approval. Publicly expressed desires were inevitably incomplete. The desire to put men to work was the immediate motivation to build the first dams, but there were other ways of putting men to work. Dams were built because they tied in with other desires.

We turn to the other reasons for big dams, opportunity and means. Opportunity in the Pacific Northwest arose from accidents of its settlement and its geology and from the legal framework for the hydroelectric development of the Columbia Basin; and the means to take advantage of the opportunity came from outside. The region lacks the usual sources of energy, oil and natural gas, and its coal is of poor grade. What it does have in abundance is water power; this resource more than compensates for its lack of mineral fuels, as the account in this book makes evident.

Many big rivers of the world are densely populated, but the Columbia River is not one of them. The West was settled fast, and railroads and highways became more important traffic routes than a river clogged with rapids. There are only four cities on this big river. One is in northern Washington, Wenatchee; one is in central Washington, the Tri-Cities, a cobbling of three adjacent towns, which owes its standing as a city somewhat to its peculiar industry, plutonium production; and two are near the mouth, across the river from one another, Vancouver and Portland. Portland is the only one of the four that is large, and it belongs more to the Willamette River, which runs through it, than to the Columbia, which runs north of it. What this implies for big dams on the Columbia is that they can be built without having to resettle large populations, as has happened elsewhere, with major social, economic, and political problems.

The typical big dam in the Columbia Basin was built for multiple uses, including hydropower. The Corps had an early brush with hydropower in 1909-17, when Congress directed it to raise the height of a low dam on the Upper Mississippi to accommodate hydropower as well as navigation; the Corps built the base for a power station, but it could not build the station itself. The Corps' Wilson Dam was a multipurpose hydroelectric dam, but it was a wartime anomaly. In 1920, a Water Power Act defined a role for the federal government in hydroelectricity, but it did not coordinate it with other river uses. In the same year, the Corps undertook a survey of the Tennessee River Basin, looking into all of its possibilities. The findings impressed Congress, and in 1924 it proposed that like surveys be taken of other river basins. The result was the act of 1925, above, directing the Corps to estimate the cost of surveying the nation's major rivers. By this time, Congress had already enacted the Colorado River Compact leading to Boulder Dam, its first significant action on the Progressive's prewar plan to maximize the usefulness of America's watersheds.

In 1926 the Corps submitted to Congress an estimate of the cost of surveys it proposed to conduct on 180 navigable rivers with hydroelectric potential, together with their tributaries; this became House document 308, the original "308 Report." Congress approved the funding the next year. Reports for twenty-four separate surveys of waterways, called the "308 Reports" after the original House document, were submitted to Congress beginning in 1930. These reports had far-reaching consequences for American rivers, and they changed the mission of the Corps of Engineers. When Theodore Roosevelt had proposed multiple uses of American rivers, the Corps had objected on the grounds that the right use of rivers was navigation, and that other uses would interfere, as they sometimes would. With its 308 Reports, the Corps dropped its objection; the reports would place the Corps at the center of basin-wide projects for the multiple uses of rivers over the next decades. These projects included major dam constructions on the Tennessee River, the Sacramento River in California, the Colorado-Big Thompson Rivers in Colorado, the Missouri River in the Great Plains, and the Columbia River.

The exhaustive 1,845-page 308 Report "Columbia River and Minor Tributaries," which the Corps submitted to Congress in September 1931, proposed ten multipurpose dams in the Columbia Basin and several upstream storage projects. Conscious of the daunting scope of the project, the authors of the report noted that these dams were "all on a large scale, some on a grand scale." The conditions of their design and construction were not ordinary either; their foundations and the floods they would need to pass were "unprecedented." However, there was nothing "to cause a belief that the engineering difficulties cannot be surmounted." The Corps, as it turned out, was willing and able to build these dams when the time came.

The 308 Report concluded that flood control was of minor concern; that irrigation was uneconomical unless combined with power development; and that improvements in navigation were served by projects already approved. It was to the remaining use of multipurpose dams, power, that the Report gave special attention, consistent with its finding that the Columbia Basin has the greatest hydroelectric power potential of any river system in the nation.

The Corps estimated the total power capacity of the Columbia Basin at 8,000,000 kilowatts, and it projected that the power generated at Grand Coulee Dam alone would meet the likely increase in regional needs for the next thirty years. These estimates would be superseded, but the great potential for hydroelectric power development in the Pacific Northwest was a reality. The Corps found the power development of the Columbia Basin economical as well as feasible, but the development would have to be incremental, with consideration given to the entire power industry of the region, lest the power exceed the demand.

The reason for the Corps' caution in recommending action on hydroelectric dams is evident from the figures it worked up for the power potential but also from the cost. The cost of building nine dams – one of the ten dams was already under construction by a utility – "will exceed that of any other single development of any kind for power that has ever been made." If the dams were going to be built, the Corps assumed that the financing would come from local authorities or private interests, the conventional thinking of the time, though on the Middle Columbia, where the Corps foresaw improvement in navigation in connection with any power development, it recommended federal financing of the entire cost of the improvement. The Corps was partially wrong about the financing of the dams, but not about the plan. The 308 Report would guide the development of the Columbia Basin over the next forty years, and the Northwest would become rich in hydroelectric power as a result.

Through the action of the federal government, the Columbia River delivered on its promise. The three largest producers of hydroelectric power in the

United States are all Columbia River dams: Grand Coulee, Chief Joseph, and John Day, and the sixth largest producer is another dam on the river, The Dalles. No other river in the country is in its class. The dams fully justified one of the motivations behind the development of the river basin, the desire for power.

THE FEDERAL DAM-BUILDERS

The opportunity, means, and desires to build federal dams in the Columbia Basin were channeled through two agencies, the Bureau of Reclamation and the Army Corps of Engineers. The Bureau, originally the Reclamation Service, was created in 1902 by the National Reclamation Act, within the US Geological Survey. A strong supporter of the act, Theodore Roosevelt wanted to turn the Western desert into a garden, in the imagery of the time, and he looked to engineering and to dams to do it, and to the federal government to support it, since reclamation was costly. The new agency was expected to finance its irrigation dams from proceeds of sales of federal Western lands, and farmers receiving benefits were expected to repay the government in annual installments. The Bureau started out with simple irrigation works involving diversion dams and canals, but before long it took on a series of projects calling for large dams. Then for some years it languished, the West having insufficient political clout to move costly irrigation projects through Congress, and a succession of presidents disapproved in principle of government interference in the economy. The Bureau's activities were further curtailed by the agricultural depression of the 1920s. To get started again, the Bureau needed an ambitious project, and it looked hopefully to either the Columbia or the Colorado River. It settled on the Colorado – Hoover Dam was its project – but it kept a watch on the Columbia too. Before long, more big projects would come its way owing to a change in administrations. Under Franklin Roosevelt and continuing under Truman, a series of comprehensive river-basin bills authorized dozens of dams, canals, and other projects on major rivers of the West. Money was little object, since the projects were expected to pay for themselves eventually. The Bureau was set on a course that would transform the arid West, physically, economically, and demographically.

The Army Corps of Engineers is much older than the Bureau. George Washington appointed the first army engineer in 1775, and the Army established the Corps of Engineers as a separate branch in 1802. From running a military academy and building forts, its activities expanded to include civil undertakings, notably river navigation in the nineteenth century and flood control in the early twentieth century. What the Corps does today is a continuation: it dams rivers, builds bridges over them, dredges and straightens them, rip-raps them, builds levies, dikes, and reservoirs on them, builds locks and canals, and builds fish hatcheries, and it maintains and operates the facilities it builds. The Corps has altered American rivers as profoundly as the Bureau.

The Corps is a hugely successful organization. With its 35,000 civilian and 650 military personnel, it is the world's biggest public engineering and construction management operation. Although in many respects, it is a civilian agency, its leadership is military. Its head is chief of engineers, who is a staff officer in the Pentagon. The military tradition from which the Corps arose and to which it is still joined is evident in the Corps emphasis on order, system, and scientific control. Its bigger-is-better approach to projects comes out of this tradition. In reading this book, it is well to keep in mind that the Corps' history and culture as well as its mission differ from the Bureau's.

From the beginning, the Bureau of Reclamation's territory was the West. Although the Corps of Engineers (together with the Corps of Topological Engineers) did much of the mapping of the West in the nineteenth century, until the Great Depression it largely worked on projects in the Middle West and on the East Coast. With the arrival of the free-spending New Deal, the Corps expanded its range of operations in the West, building a good number of dams there. The West being mostly arid and open to irrigation projects, the Bureau built more dams than the Corps, but in the Columbia Basin where there is an abundance of water the Corps led the way. Roughly two thirds of the federal dams in the basin are operated by the Corps.

With their separate responsibilities, the Bureau's for irrigation and the Corps' for navigation and flood control, their activities conveniently complemented one another, but with the advent of multipurpose dams their division became blurred. The two agencies often found themselves working on the same waterways and competing for funding. Their projects overlapped. The Flood Control Act of 1936, the legal origin of much of the Corps' work in the West, authorized the Corps to build dams to control floods, but the impounded water of these same dams could be made available for irrigation, normally the Bureau's business. For the same reason, their projects conflicted. The Corps might want all the water it could get to develop navigation downstream, while the Bureau might want the same water for irrigation upstream. In the Southwest and in the Missouri Basin the Corps and the Bureau regularly fought over water projects, jealous that the other might get them. Their enemies called them opportunistic, wasteful of public money, out only to satisfy their ambitions and to please the powers at be. Their turf wars gave water development a bad reputation.

Pork-barrel contributed to that reputation. Theoretically, the Corps – and the Bureau too, but we take the Corps as our example here – only informs

Congress about waterways, and Congress decides, but things do not exactly work that way. The Corps goes after projects it recommends, making friends with lobbying interests, and building support in Congress. The Corps and committees in Congress have a cozy relationship. People from the Corps or a committee might approach a freshman politician and ask him if his district has a water project. Why not? Then some important bill will be sent through with a rider authorizing this politician's pet dam. This is how they trade favors. The dam might not be particularly needed, but money goes to the district all the same, votes go to the politician, and jobs go to the agency. Everyone is happy but the tax payer, and today the environmentalist. That is the way pork barrel works, as most people know. The Corps gets raked over the coals from time to time, but the practice goes on. The Corps is a big institution with built-in goals and pride of achievement, and its ambition for projects is in keeping.

It is in the Pacific Northwest that the Corps and the Bureau have shown themselves to best advantage. The division of the spoils there was not equitable, the Corps receiving the lion's share, but it was not bitterly contested by the Bureau either, though their rivalries caused delays. Following a big flood in 1948, Truman ordered the two agencies to coordinate their reports on the Columbia Basin. They complied, all the more readily because the idea of a feared TVA-like agency for the region was still alive and backed by the president. They divided their responsibilities in the usual way. The Corps would get navigation and flood-control projects, the Bureau irrigation projects. As for multipurpose projects, they divided up the rivers. The Corps would get the Columbia River below Grand Coulee Dam and the Lower Snake River. The Bureau would get the Middle and Upper Snake River and most of its tributaries. In this way, amicable cooperation between the two federal dam-building agencies came about, and the Pacific Northwest was the better served for it.

We note a third large group of (non-federal) dam-builders, the hundreds of engineers employed by the TVA, who built dams on the Tennessee River at about the same time that engineers of the Corps and Bureau of Reclamation built dams on the Columbia and Snake Rivers. TVA dams, like their Western counterparts, were intended for multiple purposes, and during World War II they too supplied hydroelectricity for aluminum extraction and the production of nuclear explosive. TVA dams are important in the history of hydroelectricity and dam engineering in the US, but they lie outside the scope of the present book. The TVA as a model agency for the Columbia Basin, however, will reappear in the section on marketing power.

That federal agencies, the Bureau and the Corps, led the hydroelectric development of the Columbia Basin should not surprise us in light of the history of the American West. The federal government took an active role in its settlement, providing land for homesteading, grazing, timber, and schools, facilities for navigation and flood protection, incentives for railroads, soldiers for maintaining order, engineers for mapping, researchers for agriculture, and more. The history of the American West, as historians have recognized for some time, is not only, and not primarily, the colorful history of fur trappers, gold miners, and sheriffs; among other things, it is the history of federal government, bureaucracy, power, money, technology, and experts. In light of the importance of water and its control in this history, an eminent historian has called the American West an "hydraulic society." If we carry this through, we see in Mac, a federal civil servant and expert in hydraulic design, a personification of the American West in the twentieth century.

FIRST FEDERAL DAMS

We begin this section on federal dams with a historic photograph of a symbolic meeting of our two kinds of power in the Pacific Northwest, human and physical. It shows the president of the United States Franklin Roosevelt at the dedication of the first federal dam to be completed on the Columbia River. The following discussion tells how this meeting came to pass.

10. Franklin Roosevelt at Bonneville Dam, 1937. The president approved, funded, and dedicated Bonneville Dam, all nearly within his first administration. Bonneville Power Administration.

The 308 Report projected that of the total Columbia River head of 1,243 feet – the drop in elevation of the river from the international boundary with Canada to the sea – all but ninety-five feet would be used to generate electric power. The fully developed Columbia River outlined in the Report was nearly realized. What the Report could not foresee was the need for another kind of federal power for which there was a more urgent desire, nuclear power. This need came about when the Columbia River was selected as the site for plutonium production for atomic bombs in World War II, the Hanford Nuclear Reservation. The plutonium plant needed huge quantities of water and elec-

11. Hanford. In the foreground we see the New Production Reactor at the Hanford site on the Columbia River. Unlike the previous reactors, which had a single purpose, the production of plutonium for weapons, this reactor was built both to produce plutonium and to generate power for the Northwest Power Pool through the Washington Public Power Supply System. To make up for power shortages until new dams came on line, Congress in 1962 authorized the Bonneville Power Administration and the Atomic Energy Commission to enter into financial agreements with sixteen public utility districts in Washington State to build and operate the multipurpose reactor. The physics of it was unexceptionable – steam from the reactor, otherwise wasted, would generate electric power – but the plan was ill-conceived. The BPA overestimated the future rate of increase in regional power demand and the attractiveness of nuclear thermal power plants. It encouraged the public utilities to commit to the system, and when the system defaulted on bonds, its rate-payers were the big losers. The New Production Reactor was closed in 1987. Today the system, renamed Energy Northwest, operates one nuclear plant at Hanford. Plumes from three other reactors are visible in the distance in the photograph. PD-USGOV.

tricity, but for security reasons this reach of the river was left undeveloped. For this, the last natural haven of the river's salmon, we have the ultimate weapon of human destruction to thank; our hunger for power makes for strange companions at times. A photograph of a Hanford nuclear reactor is included in this book for completeness, showing another way the river is made to serve us in the name of power.

When the Corps of Engineers undertook its survey of American watersheds, it was understood that neither the Corps of Engineers nor the Bureau of Reclamation could build dams primarily for the purpose of power; the Corps naturally expected that the construction of hydroelectric dams on the Columbia River would be a business venture or a state undertaking, and the first dam supported the expectation. It was Rock Island Dam near Wenatchee, Washington, constructed by the Puget Sound Power & Light Company between 1930 and 1933. To build a dam on a river that size was a considerable engineer-

12. Rock Island Dam. The original powerhouse was expanded and a second powerhouse was added. This run-of-the-river power dam is now owned by the Chelan County Public Utility District. Northwest Power and Conservation Council.

ing accomplishment, but the dam itself was a minimal dam. It incorporated fish ladders because it was required to, but it did not have a navigation lock, and it did not provide for irrigation or flood control. It was intended solely for power. Later several more dams on the Columbia were built by power and light utilities, with the same narrow intention. By contrast with these single-purpose dams, the federal dams on the Columbia, beginning with Bonneville, were built for navigation improvement, irrigation, and flood control, as well as for power.

Franklin Roosevelt was elected president soon after the Corps submitted its report on the Columbia waterway to Congress, and he brought with him to Washington a different understanding of hydroelectric power and government. He wanted electric power to do more than merely offset the costs of federal multipurpose dams, maybe with a little surplus left over. He wanted it to be the "servant of the American people," supplying wide domestic and rural markets. On the campaign trail, he had spoken of the "vast possibilities of power development on the Columbia River"; long a champion of federal electric power on the Niagara and St. Lawrence Rivers in his own state, New York, and an admirer of Wilson Dam on the Tennessee River and the Boulder Project on the Colorado River, he had stated "in definite and certain terms, that the next great hydroelectric development undertaken by the Federal Government must be that on the Columbia River." In his first year in office he followed through; guided by the Corps of Engineers' recommendations in its 308 Report on the Columbia River, he approved and funded the first two federal dams on the river, Bonneville and Grand Coulee. As with other New Deal projects, he got around the need for Congressional approval of the dams by using his authority to fund emergency relief as Public Works Administration projects under the National Industrial Recovery Act. Following a Supreme Court decision in 1935, Roosevelt had to go through Congress for approval, but the jump start from relief funds left Congress little choice but to authorize and fund dams already well under way.

Through the power of the purse, Roosevelt acted resolutely to establish federal authority over Western river resources. In the standoff between private and public ownership of hydroelectric power, he came down squarely on the side of public power, and in a contest between local and federal authorities he favored federal authority. There was a party that wanted the state of Washington to build the dam at Grand Coulee, but that would have entailed a delay in the courts over land issues. Under US statutes, the federal government could avoid this; Roosevelt proceeded with a federal dam without delay. It speaks of his belief in federal electric power that the public works money he drew on to start up Grand Coulee was the largest sum allocated for any purpose under the National Industrial Recovery Act.

Within the federal government, there was a question of which agency would build the dams. The second of the two dams to be authorized, Bonneville, was handed to the Corps of Engineers, which was deserving, having long been active on the Lower Columbia River. The first dam, Grand Coulee, was contested. The Corps had reversed itself on its recommendation of a high dam at Grand Coulee in the 308 Report, but when prospects for the development of the site looked up, it reversed itself again, proposing to build a dam there, this time a low dam to produce power. In the end, the department of interior gave the job to the Bureau of Reclamation. The Corps lost out in part due to its vacillation – it would not hesitate again, it would build all of the other federal dams on the Columbia River – but the Bureau had a good case. The Grand Coulee project had long been associated with irrigation, the Bureau's mission, and the Bureau's stock as a big-dam builder had never been higher. The same year that Grand Coulee was approved, the Bureau began pouring concrete for Boulder/Hoover Dam, which when it was completed three years later would be the biggest dam in the world, and an engineering triumph.

Critics of Roosevelt's dams called them a waste of money and a fraud on the taxpayer, but once they were started their completion was never in doubt. What began in 1933 as a relief measure in the depths of the Great Depression was finished on the eve of World War II as a defense project. The press was largely enthusiastic, and the crowds of tourists found the dams an antidote to

13. Grand Coulee. This canyon, or coulee, is the ancient river bed from which the dam takes its name. It lies about 600 feet above the present river bed, and its walls reach 1,300 feet or more. The photograph was taken at a scenic viewpoint by the author.

14. Grand Coulee Dam. This first federal dam to be authorized on the Columbia River, Grand Coulee Dam, is located 450 miles upstream from Bonneville Dam, the second dam to be authorized and the first to be finished. Grand Coulee Dam went into operation in 1941, the year before this photograph was taken, when the dam was completed. Water is pumped 280 feet from Lake Roosevelt, the reservoir behind the dam, into Banks Lake in the upper Grand Coulee, the principal reservoir. Banks Lake supplies a vast irrigation network, and it is also used to boost and steady the hydroelectric power generated at the dam; water is pumped into it from Lake Roosevelt when power demand is low and is released to fall back to Lake Roosevelt through a powerhouse when demand is high. The dam was spilling water at the time of this photograph, an impressive and somewhat uncommon sight. Albert L. McCormmach.

despair. That such projects were undertaken in the midst of the greatest economic catastrophe in the country's history was proof for them that the country would endure and recover.

When Bonneville Dam was finished in 1937, Roosevelt naturally came out to give the dedication speech. He assured the people that this great effort was going to pay off, that it was going to bring them wealth and the "widest possible use of electricity." He brushed off the criticism that the dam was a white elephant because there was no market for the electricity, confident that once the dam was in business, it would prove him right, as it did. When President Truman gave the delayed dedication speech at Grand Coulee Dam in 1950, he too brushed off similar criticism, now with ample proof an ever growing demand for power. Grand Coulee Dam, he said, had "transformed the energy of a mighty river into a great new source of national strength"; criticism of this dam, it seems, was by then unpatriotic as well as undeserved.

Whatever the other benefits of Bonneville and Grand Coulee, the most far-reaching was the electric power they delivered. It lighted rural homes, lessened the drudgery of farm labor, and ran factory motors. Telephones and radios became commonplace. Flatirons were replaced by electric irons, wood stoves by electric stoves, wash boards by electric washers, ice boxes by refrigera-

15. Bonneville Dam. Built by the Army Corps of Engineers in 1933-37, this first federal dam on the Columbia River to go on line is located at tide water, 140 miles from the Pacific Ocean. It is built in several sections, as the aerial photograph shows. The island on the left was created in the 1970s when the river channel was widened to accommodate a second powerhouse. US Army Corps of Engineers.

16. Bonneville Dam. This photograph taken around 1940 shows water coming over the spillway of the dam, and in the foreground we see a section of the Oregon shore fish ladder. Albert L. McCormmach.

tors, brooms by vacuum cleaners. The federal development of hydroelectric power together with other New Deal measures transformed the factory, the farm, and the home in the Pacific Northwest, altogether a regional revolution.

MARKETING POWER

Northwest conservatives wanted to limit the role of the federal government in the development of the Columbia River Basin. Private enterprise, they argued, should be allowed to capitalize on any economic opportunities arising from it. Northwest progressives wanted public ownership of hydroelectric power, making the case that since electricity had become an essential fact of life in modern America, the public deserved to be protected from private exploitation. They wanted a regional authority.

When the New Deal arrived, the idea of regional planning was big, and hopes for it ran high. Early on it produced the Tennessee Valley Authority, the New Deal's most successful, and most controversial, regional agency. Approved the same year as Grand Coulee and Bonneville Dams, the TVA set about to transform the Tennessee River into a series of working reservoirs, following the recommendations of the Corps of Engineers' report on the Tennessee River. The TVA's secret weapon in the war against the Great Depression was multipurpose dams, which supplied the region with copious, cheap electricity. (Electricity proved the easy part; some of the TVA's social goals went unrealized.)

Roosevelt viewed the TVA as a model agency for the conservation of natural resources in the nation, and in 1937 he asked Congress to create a similar authority in seven other watersheds, including the Columbia. In its full version, a Columbia Valley Authority, or CVA, would have had the power to coordinate the generation, distribution, and sale of electric power, the reclamation of lands, the creation and operation of agricultural and industrial enterprises, the management of fisheries, and the planning of cities throughout the Columbia Basin. Between the mid-1930s and the mid-1950s, legislation to create a CVA was proposed repeatedly, with steadily weakened variants. The dream of a CVA died hard, but it probably never had a chance. To most natives of the Northwest, with their ingrained suspicion of big government, a Columbia River version of the TVA sounded like socialism. The benefits seemed to them dubious at best. The Northwest after all was not an old, worn-out land of sharecroppers and subsistence farmers like the South; it was a new, rich land of independent farmers and entrepreneurs. Irrigation interests, businesses of all sorts, political groups, private utilities, and even public utilities all wanted to hold onto what control they had, not forfeit it to a CVA. Further, the Corps of Engineers and the Bureau of Reclamation were by then both entrenched in the West, and these agencies were conservative, generally aligned with opposition to federally driven social change; disagreeing as they often did about projects, the Corps and the Bureau agreed that a CVA was unnecessary for the development of the Columbia Basin. The BPA (discussed below) was a proven success, a decisive point. The final nails in the CVA coffin were the distractions of the Korean War and the McCarthy witch-hunts, which made every kind of social planning politically suspect.

People living in the major river basins during the New Deal – the Colorado, Tennessee, Missouri, Central Valley, and Columbia basins, prominent among them – worked out their own forms of organization and technology, each different. Regions had their own physical, social, economic, and political characteristics, which affected the way the federal investment in their rivers played out. To be sure, the ambitions and conflicts of the Bureau and the Corps, the administration, Congress, and local governments reappeared in each setting, but never quite the same way. In keeping with how changes come about in America, the Northwest chose an alternative both to a CVA and to private corporations for the production and distribution of Columbia hydroelectricity. The region wanted the federal government to build big multipurpose dams and to create a central agency to manage the hydroelectricity they produced but to leave everything else alone.

While Bonneville Dam was still under construction, there was a heated argument over its electricity, who would sell it and how it would be marketed. Some parties wanted the Corps of Engineers to run the dam and to sell its power at the dam site on a come-and-get-it basis. Others wanted a more CVA-type of arrangement. The Oregon State Planning Board wanted the electricity priced at a variable rate depending on distance, the cheapest rate close to the dam, to attract industries to Oregon and end its colonial status as a raw materials state. The arguments against this were that it would do nothing for rural electrification and domestic access, it would interfere with an anticipated interconnection with power from Grand Coulee Dam, and it would fill up the scenic Columbia River Gorge with heavy industries; the arguments prevailed. Other groups with powerful political backing wanted a uniform electric rate in the interest of domestic users, the administration's preference, and the one which was adopted. When power from Bonneville Dam was ready to go on line, Roosevelt ordered the secretary of the interior to prepare legislation to handle it. The result was the Bonneville Project Act in 1937, which established an independent agency of the interior department responsible for marketing hydroelectric power from Bonneville Dam. To stimulate the development of the entire region and to facilitate rural electrification, the Bonneville Project imposed uniform electric rates, set by the cost of generation and transmission.

They were called "postage-stamp" rates, modeled on single-price postal stamps, good for sending mail across the street or across the nation. The object of the new agency was to insure the "widest possible use of available electric energy," which of course was Roosevelt's policy. The agency was to prove highly successful in realizing its object through the 1940s and 50s.

Marketing implies a market, and when Bonneville Dam was built there was an insufficient market for the power, so part of the new agency's job was to promote its product. In this respect, the Bonneville Project was an active sales operation. With its creation, the federal government overcame its traditional reluctance to go into the power business.

To protect federal power from domination by special interests, the new agency was required to give preference to publicly and cooperatively owned utility districts, which were already strong in parts of the Northwest and which soon multiplied as a result of the agency. The purpose was to restrain private enterprise, not to replace it. The agency did not take over private utilities, and it delivered power not directly to consumers (with the exception of a few large industries) but to private and public utilities, municipalities, and rural cooperatives, which were free to sell it, prevented only from overcharging. The utilities were allowed to build and operate dams, but the federal government set the electric rates. The builders of the federal dams, the Corps of Engineers and Bureau of Reclamation, were allowed to operate their dams too, again with rates set by the new agency. Bonneville Project's postage-stamp rates and its encouragement of public utilities constituted a social policy, but not socialism. In contests over the rates the agency set for the Columbia Basin, there were now three principal players, the Bonneville Project in addition to the Corps of Engineers and the Bureau of Reclamation. (From the start, the Project and the Bureau were on a collision course: the Project wanted to keep electrical rates as low as possible, and the Bureau wanted them high to pay the costs of irrigation.)

The immediate purpose of the Bonneville Project was to market power from Bonneville Dam, but it or its successor was expected in time also to market power from Grand Coulee and from later federal dams. This intention was incorporated in the planning from the start. In 1940, by executive order, Roosevelt formally expanded the Project's responsibility to include power from Grand Coulee. To insure the dependability of the power supply everywhere, a study commissioned by Roosevelt in 1936 recommended an interconnection of long-distance lines from Bonneville and Grand Coulee Dams and also from future federal dams. The first interconnection was a line from Grand Coulee Dam to the substation on the Washington side of Bonneville Dam in 1940, the year before the first unit at Grand Coulee went on line. Coordinated in this way from the start, the electric power coming off the two dams was the world's greatest concentration, and the Bonneville Project was in charge of it all.

In 1937, the Project became its own successor. (When Roosevelt signed the Bonneville Project Act, he hoped it would be replaced later by a more comprehensive CVA, which was not to be.) The name of the agency was changed to Bonneville Power Administration or, as it came universally to be called, its acronym, BPA. The name, we see, retained "Bonneville," which was now somewhat misleading, but which implied continuity; perhaps more significant, the name contained "Administration" rather than "Authority," which was associated with the rejected "Columbia Valley Authority" and its inspiration the "Tennessee Valley Authority." In 1942 the War Production Board required the BPA and the ten major utility systems using BPA power lines to enter into a forum for coordinating operations, the Northwest Power Pool. There were already interconnections between Northwest utilities going back to the 1920s, and the BPA had interconnections with Portland General Electric Company and the Cities of Seattle and Tacoma, but only Uncle Sam could make a fully integrated system of power lines for the region. Today the power pool, one of ten electric power markets in the continental US, covers part or all of eight states, an updated, enlarged, quasi-giant power. The interconnection was taken a step further in 1964 with the Pacific Northwest Coordination Agreement, by which the BPA, the Corps, utilities, and, later, the Bureau agreed to operate the dams as if they were owned by a single body, optimizing the production of power. The entirety of BPA lines and hydroelectric dams built by the Corps and the Bureau and operated as an integrated system is known today as the Federal Columbia River Power System (FCRPS).

The desire for some version of giant power continues. The time of big-dam building in America may be history, but the networks of power lines that were built at the time of the dams are still in place and expanding. Like obsolete dams, they can be decommissioned, but if they are, they are rebuilt as long as there is a use. One of the proposals to counter global warming is to replace the current, antiquated regional patchwork of power-line ties with a single, efficient, nation-wide network of underground power lines. To stimulate the economy and at the same time increase energy efficiency, the Obama Administration made available grants and loans to modernize the electric power grid (Appendix B).

POWER LINES

With the preliminaries now behind us – a dam ready to produce power, Bonneville, and an agency in place to market its power, the BPA – we come to our mid-level federal civil servant, Mac. To carry the electricity from Bonneville Dam to its users, long-distance power lines had to be built. These were what got Mac started on a course that led him to a career in the Army Corps of Engineers as a designer of more dams like Bonneville that would produce more power to run on the BPA lines.

Since Mac is our guide through the maze of power lines that march in file across the Pacific Northwest, we ought to know what he looks like. Spread before me are a number of photographs taken by my mother or by me. One is slightly out focus, a telling detail. Photographic sessions in my family were tense, and my anxious mother had wiggled the camera that time. Mac is wearing a short-sleeve shirt, showing his well-muscled arms. Six feet even, weighing 175, there is nothing soft about him anywhere. He gives the impression of physical power beyond the ordinary. I have several photographs taken of Mac near the breakup of his marriage; in these, his eyes are narrowed, like a fighter's. Photographs from his working years show a lean, prominent jaw, thrusting chin, high cheekbones, and dominating nose. The forehead is broad, the chin is squared off, the eyes are blue and wide set, the skin across the forehead is pulled tight, as though stretched to the limit, ready to snap. Mac can frighten people. I also have several photographs showing him at unguarded moments and others showing him in professional groups, seemingly at ease with the world. In one, he is wearing a checkered jacket, a wide shirt collar spread over its lapels, the sporty look of the time. His early receding hairline is not obvious, since barbers left little hair on male heads in those days. I have a photograph of Mac near the end of his life. He stands scarcely more than half my height, bent double by Paget's disease. Standing tires him, since the head is a heavy object to hold up when it is not balanced. Even though it is a hot day, he wears a long-sleeved shirt to conceal his once powerful arms, now shrunken and stringy. The ravages of age have scaled down this dam-builder, but his eyes are as expressive as ever. He does not miss a thing.

To understand what brought Mac and power lines together, we need to know how he came by his skills and opportunities. Mac grew up in the dry-farming, wheat country of eastern Oregon. His hometown Pendleton is known for its annual rodeo, the Pendleton Round-Up, and for the Pendleton Woolen Mills, which today serves a worldwide market, everywhere recognized by the original Indian designs it used from the start. Mac's father ran the first automobile business in this cowboy and Indian town, and at the same time he owned and operated a wheat ranch not far from town. He was a high-energy, restive man. On one hot summer day, in a state of physical exhaustion brought on by overwork, he came down with pneumonia, and in no time he was dead. He had not planned to die, and he did not leave his large family with much to live on. Left to his own devices, Mac enrolled at the state college in Corvallis, where he took courses in agriculture and mechanical engineering and supported himself by cutting firewood and doing odd chores for sororities. His professors thought that eastern Oregon farmers were mechanical geniuses, because they knew how to adapt their machinery to work their enormous wheat fields. The professors were not interested in that kind of farming themselves but in growing all sorts of things and raising all sorts of animals. Mac was bored by his courses, and short of money and lacking direction he dropped out.

Upon returning to Pendleton, he noticed some construction underway.

17. Mac, Late 1930s. He was a new employee of the Bonneville Project when this photograph was taken in his rented house in Portland. Albert L. McCormmach.

He asked for a job and was handed a shovel. He got down in a pit to dig holes for the piers of a building, for which job he was paid the going rate for laborers, fifty cents an hour, four dollars a day. After a few weeks, when a shipment of steel was delayed, everyone was laid off. Mac did not wait for the steel to arrive, but walked down the street until he came to the Pendleton Roller Mills, where he was hired at the same pay. Wheat flour arrived at the mills in boxcars with openings on the side for unloading. Wheat that did not flow out of the openings had to be shoveled out, and Mac was again the man with the shovel. This was dusty work, and Mac suffered from hay fever. One day he was too ill to go to work, the next day he did not have a job. He walked down the street again, stopping at an old building with a large hall for lodge meetings, where a little hand-printed sign caught his attention: "Oregon State Highway Commission – local office upstairs." He went upstairs and asked about a job and was told to write his name at the bottom of a long list of applicants. A few days later he was notified that he could work for three days. At the end of the third day, nothing was said. He worked a fourth day, a fifth, and when he asked, he was told he could keep coming back until he was told to stop. The Highway Department had a big consignment of crushed rock, which had been dumped at various locations, and Mac's first assignment was to compute the quantity of rock. He measured cross sections of the piles and made calculations with a Monroe mechanical hand calculator. He had taken a course in surveying in college, and he was good at mathematics, so the highway people thought they had found a gem in him. Soon he was making $100 a month.

Then one day he met a man who said he was quitting his job as draftsman at a local title company, Hartman Abstract. The salary with Christmas bonus came to $150, one and a half times what the highway department paid. Mac applied and was hired. This was at the time of the agricultural depression, soon to be followed by the Great Depression, and with all of the forfeitures going on then, the title company had a brisk business. Through nine years of the nation's economic hardships, he worked on land titles and legal descriptions. With a steady job, he was one of the lucky. He married; he and his wife lived in a tiny, rented house and had a healthy little boy in the shape of a perfect cube. That boy was me.

While walking through the Pendleton post office one day, Mac's eye fell on a notice pinned to the bulletin board announcing a federal civil service examination for various kinds of draftsmen. Curious to see if he remembered anything from school, Mac took the examination. In a couple of weeks, he received a rating in the mail for three kinds of draftsmen. It was 1938, the year Congress approved funds for the Bonneville Project to begin building a network of power lines from the just-completed Bonneville Dam, and the agency

18. Title Company, Late 1930s. This shows two of Mac's coworkers in an idle moment gazing out the window of the Hartman Abstract Company in downtown Pendleton, Oregon. An employee there for nine years, Mac seemed headed for a career in the title business. Albert L. McCormmach.

was looking for draftsmen. No sooner had Mac been notified of his rating than he received a telegram from the Bonneville Project office in Portland, which had a register of names from the ratings, offering him a job as a topographical draftsman. He had not been surprised when he first learned about Bonneville and Grand Coulee Dams, since as a schoolboy he had often heard that the Columbia River would be developed as a power stream one day. A glance at its rapids told why. Most big rivers do not have rapids but are flat, like the Yakima River. By contrast, the Columbia River is both big and steep, having volume of flow and drop, both, which are what make it ideal for producing power. Big as the Columbia is, it has a power potential even greater in proportion to its size.

Even though the Bonneville Project salary was exactly what Mac earned at the title company, $1800, he figured that by knowing a little about surveying from a college course and from his work for the highway department and by having experience with rights-of-way and easements from his work for the title company, he was just the kind of man the Bonneville Project was looking for, and that in a few years he would be making good money. He quit his job at the title company and brought his family – still a prudent Depression-era family of three – to Portland.

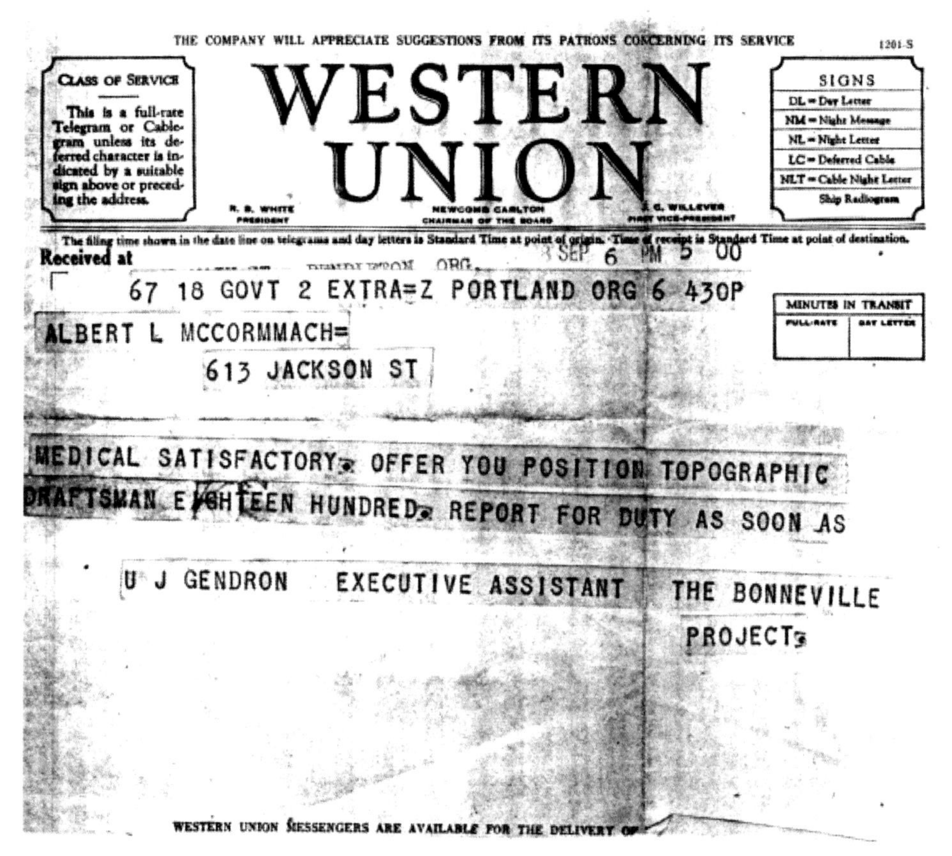
THE COMPANY WILL APPRECIATE SUGGESTIONS FROM ITS PATRONS CONCERNING ITS SERVICE 1201-S

CLASS OF SERVICE
This is a full-rate Telegram or Cablegram unless its deferred character is indicated by a suitable sign above or preceding the address.

WESTERN UNION

R. B. WHITE, PRESIDENT — NEWCOMB CARLTON, CHAIRMAN OF THE BOARD — J. C. WILLEVER, FIRST VICE-PRESIDENT

SIGNS
DL = Day Letter
NM = Night Message
NL = Night Letter
LC = Deferred Cable
NLT = Cable Night Letter
Ship Radiogram

The filing time shown in the date line on telegrams and day letters is Standard Time at point of origin. Time of receipt is Standard Time at point of destination.

Received at SEP 6 PM 5 00

67 18 GOVT 2 EXTRA=Z PORTLAND ORG 6 430P

MINUTES IN TRANSIT — FULL-RATE — DAY LETTER

ALBERT L MCCORMMACH=
613 JACKSON ST

MEDICAL SATISFACTORY. OFFER YOU POSITION TOPOGRAPHIC DRAFTSMAN EIGHTEEN HUNDRED. REPORT FOR DUTY AS SOON AS

U J GENDRON EXECUTIVE ASSISTANT THE BONNEVILLE PROJECT.

WESTERN UNION MESSENGERS ARE AVAILABLE FOR THE DELIVERY OF

19. Telegram, 1938. In August 1937, Congress passed the act that established the Bonneville Project. A year later Mac received this telegram, which changed the direction of his life. The missing word in the telegram is "possible." Bonneville Dam had just gone on line, and the Bonneville Project was in a hurry to put up long-distance power lines. Albert L. McCormmach.

The Bonneville Project's first item of business was power lines, and it lost no time. It designed a transmission system, surveyed land for running the lines, acquired rights-of-way for the land, let bids, and within nine months after it was created it began construction. In 1938 it began transmitting power over its own facilities, first a three-mile line to Cascade Locks, then in 1939 a major 230-volt line to Portland General Electric Company. The Project was moving so fast on its lines that it practically lived off the street. Mac joined a large number of other new employees crowded into a room over a garage on Union Avenue, way out in north Portland. Instead of filing cabinets, they made do with shelves improvised from orange crates. Everybody had a calculator, and Mac's first job was to compute areas of land. He would make his computations, and some one else would check them; then some one else would give him his computations to check, and it went back and forth like that.

20. Bonneville Project, ca. 1939. Note the crowded tables and the mechanical calculators. Note also the formal dress. Suits, vests, and ties are the order of the day. Men are smoking pipes, no cigarettes. Albert L. McCormmach.

After a while, the agency moved into a building of its own. The new drafting room was the size of a gymnasium, with two or three hundred drafting tables, lined up in the government way, maybe eight in a line, with maybe thirty lines. (In due course, people in the drafting room would be separated into workers and bosses, and the quasi-democratic arrangement of tables was transformed into so many fiefdoms, partitioned off by separating walls. That too was how the government did things.)

Titles were important in government work. They were to Mac, as they will be to us as we follow his course from beginning draftsman to his final destination, as yet unknown to him, as government engineer. At the head of each line of draftsmen sat the man in charge of it. He was officially classified as an assistant engineer rather than as a draftsman. Seated behind the assistant engineer was the principal draftsmen, behind him the senior draftsmen, and finally, making up the rest of the line, the lowest, the beginning draftsmen like Mac. As it happened, in Mac's line there were no senior or principal draftsmen, only the head draftsman and beginners like himself. One day the head draftsman left for another job, and Mac who sat next in line began handling his office mail. At about this time, a civil-service analyst came out from Washington,

21. Bonneville Project, ca. 1939. More room. Albert L. McCormmach.

DC to see if employees matched their job descriptions and that sort of thing. He immediately spotted the irregularity. He reclassified the position of assistant engineer as chief civil engineering draftsman. That was fair, since what the head of any line was doing was only drafting after all, not engineering. Then one day, the head of the whole operation called Mac into his office and told him to make out application form 57 for the new highest rating the analyst had put in place, chief civil engineering draftsman. Since Mac was already doing the work, the application went through without a problem.

We pause here to appreciate what has happened. Normally, a person in civil service worked for three or four years at one rating, then for three or four more years at the next, and so on, but at the Bonneville Project everything was in flux, decidedly abnormal. In not much over a year, Mac had moved up three ratings, from bottom draftsman to top. It spared him years of work. (It incurred the rancor of an older draftsman in his carpool who had seniority over him but who was, by oversight or design, not considered for the position of head draftsman. When Mac mentioned his form 57 and what it was for, this man tried to take Mac's job away from him, going to the very top of the civil service organization. This was the beginning of another lesson in Mac's education as a civil servant, cut-throat office politics.) Mac had had his first break. He now sat officially at the head of his line. He was in charge, and he was rich.

At the war's end, Mac got out his crystal ball, and what he saw was power.

22. Bonneville Project, ca. 1939. Draftsmen at work in a spacious room with natural light. A typist has a desk between the drafting tables. Albert L. McCormmach.

After a late start, he found himself at age thirty-four earning the enviable income of $2600.

Word got out that Mac was a kind of expert on legal land descriptions because of his work at the title company. He was pulled out of his line of drafting tables in the engineering division and transferred to the land division. The Bonneville Power Administration – the Bonneville Project had changed its name by now – was acquiring right-of-way deeds to huge corridors of land for its power lines, and the pace was hectic. Mac dictated legal descriptions to a stenographer, who typed fifty words a minute. He sometimes had two stenographers, first loading up one, and then the other, and he put it out all the faster. That was what Mac was doing when war was declared.

23. Power Lines, ca. 1940. Bonneville Project tower. Albert L. McCormmach.

MAC, ENGINEER

With America's entry into World War II, the BPA immediately stopped work on all power lines except those that served war industries, which it stepped up. National defense had top priority, that was the way America won the war, and the Columbia River was a national resource, a powerhouse. Here numbers tell the story: starting with no power lines and the bare beginnings of generating capacity in 1939, the BPA had 2,500 miles of power lines and 1,350,000 kilowatts of generating capacity five years later. Over ninety percent of BPA power went to war production.

In the war years, the BPA employed as many as 4,000. At some point a big layoff was inevitable. Mac who saw it coming went to the man with the top secret list of names. Sure enough, his name was on it. Instead of waiting to be fired, he walked over to the US Army Corps of Engineers, which was building air fields and training stations all around the Pacific Northwest. Because men were being called up everywhere, the Corps was left short-handed, and it was hiring anyone who could sit up. Mac was asked if he wanted a raise. When he said no, eyebrows lifted. A new person normally asked for two or three steps of administrative raises, which were routinely granted. Mac, recalling what he had learned at the BPA, asked that his rating be changed from *draftsman* to assistant *engineer* at the same salary. This was the title, undeserved to be sure, that the Washington analyst had taken back at the BPA. It was wartime, again not a normal time, and the Corps granted Mac his request right off. It meant nothing to the Corps at that time, but it changed Mac's life forever. He had exchanged the immediate gratification of a pay raise for a one-chance opportunity to advance his qualifications. With this move, his sights were no longer confined to the limited horizon of draftsman. This was his second break, the

24. Bonneville Power Administration, 1942. The Bonneville Project had changed its name to the Bonneville Power Administration by the time this photograph was taken. America was at war; shown here is the BPA's Wartime Civilian Defense Organization of the Engineering Division. Mac is in the third row, fifth from the right. US Bonneville Power Administration.

25. Bonneville Power Administration, 1942. Coordinator's staff of the Civilian Defense Organization. Mac is second from the right, unsmiling as he usually is in photographs. US Bonneville Power Administration.

payoff of his increasingly savvy penetration into the fusty world of civil service rankings. The college dropout Mac was now an assistant engineer, the first step in his climb to the professional grade of associate engineer. (Nowadays, the professional distinction is dropped, a person in government service being simply a GS 7, GS 14, or whatever.) Mac's career at this point was the illegitimate issue of Lady Luck and Mars. The war put Mac's new career as an engineer on hold – he spent it pulling standard plans for military airfields – but his new title confirmed a new direction his life had taken.

The war had brought new industries to the Pacific Northwest, doubling its power usage, and – no less important – there would be no lasting letup in the

26. US Army Corps of Engineers, Early 1940s. Mac had left his job as draftsman in the BPA and was now an engineer in the Corps. Shown are employees of the Portland District of the Corps. Mac is in the second row, far left. US Army Corps of Engineers.

demand for power with the Cold War getting underway. Truman, the new president, urged Congress to appropriate funds to speed up the construction of power dams in the Northwest. Mac knew about how long it took to approve and build a dam, and he knew how many dams were planned. By simple arithmetic, he figured that when the dams were all built, he would be ready to retire. The uncanny accuracy of his calculation owed to American defense spending but also to the readiness of Congress to bankroll giant federal projects such as interstate highways in the 1950s and space missions in the 1960s.

At the end of World War II, there were still only two federal dams on the Columbia River. Eisenhower, not an exponent of federal power, wanted a "partnership" between government and private development, and four dams were built in the 1950s and 60s on the Middle Columbia River by consortiums of public and private utilities; but after a pause, under his administration federal power in the Northwest kept right on growing. The Kennedy who followed Eisenhower was a strong supporter of federal power, and under his administration the development of Northwest rivers was offered opportunities unequaled since the beginning of the New Deal. Kennedy's successor Johnson continued the same policies. The federal government came to invest more money in power in the Northwest than in any other part of the country. Mac's career as a government engineer developed in this propitious setting.

It so happened that at the end of the war, Mac was put to work on a tedious cost-benefit job having to do with floods. The history of the job went back before the war. Flood control had become a federal responsibility with the Flood Control Act of 1936, which at the same time had established the "cost-benefit ratio" as the justification for federal projects. Its virtue was its simplicity; if the benefit in dollars of a project was greater than the cost, the project could be approved, otherwise probably not. Economics was now the standard in place of the Progressive era's scientific efficiency. The same Act asked the Corps to study the Willamette River, a major tributary of the Columbia. In its 308 Report on the Columbia and its tributaries, the Corps had downplayed the threat of flooding there. It reconsidered its position in 1937, recommending seven multipurpose dams in the Willamette watershed, with an emphasis on flood control; the next year Congress authorized the dams. Work had begun on three of these dams before the war put a halt to it. After the war, work was resumed on the three dams, and in 1947 work was started on two more. Two of the original seven were not built, but plans were laid for fifteen more; today the Corps operates thirteen dams. To secure funding for any one of them, the Corps had to show that the dam was worth it, on a cost-benefit accounting, of course. This required the Corps to send interviewers around asking residents how much damage they had in their house after a flood, how much water was in their basement, questions like that, and the answers would go into reports. The idea was to come up with a curve that plotted flood elevation against the number of dollars of damage. From that, the Corps had the basis for considering building a dam to give proper flood protection, as related to cost. How much would the dam save and how much would it cost? The money the government saved by preventing flood damage had to be favorable. Mac was assigned to this work, which he hated.

Luck was with Mac once again. The man who had supervised him during the war read the handwriting. He saw that power was coming, that power studies were going to be important. When the war ended, he was put in a corner with a little group of people in the planning branch, called "Hydraulics and Power." This combination of fields correctly anticipated the Corps' long-range intentions for the Pacific Northwest. Mac's former supervisor was very smart, an engineering graduate, and a specialist in hydraulics, who had worked for the Bureau of Reclamation, which up to that time had built most of the dams. He had a good deal of experience in hydraulic design, as no one else in the Portland District of the Corps of Engineers did; he gave a little class on hydraulic design in 1944, which Mac attended. He was also a kind of specialist in power studies. One day Mac walked over to his former supervisor and told him that he could not take his boring work on Willamette Valley cost-benefits any longer. If his supervisor had said, Well, you stick with it, you'll learn, Mac's story could have ended here. But he said, Fine, we'll have you transferred over. He already had a man working on power, so he put Mac to work on the other half, hydraulics. (Mac always spoke of his supervisor – his name was Ken Tower – with gratitude.) After a time Mac had assistants working under him, at first one or two, eventually four or five. He had become the head of the hydraulic group in the hydraulic and power subsection of something else in the planning branch of the Portland District of the Pacific Northwest Division of the Army Corps of Engineers. He had found a niche, and it felt right for him.

And that was how Mac became a hydraulics engineer. At age forty, he at last had a clear direction, a career with a name. When the work load got heavy, his former supervisor considered bringing into his group a man with more experience in hydraulic design. This man would have taken over from Mac, and Mac did not want that to happen. So every time a new job came up, Mac would prepare himself ahead of time, making certain he was always ready. He burned a lot of midnight oil in his early years, learning about hydraulic design. His effort would be rewarded; although his only formal training was his supervisor's class, equivalent to two or three college credits, he would do the major hydraulic design on a number of big dams on the Columbia River and its tributaries.

MAC'S MOVE

American security and the proven value of hydroelectricity and river transport for defense in World War II led Congress in its River and Harbor Act of 1945 to authorize McNary Dam, the third federal dam on the Columbia River, for power and river development, and in addition to authorize such dams as were needed for navigation on the Lower Snake River. Chief Joseph Dam, the next dam downstream from Grand Coulee, was approved the next year. What had changed since 1937, when the Corps had considered these same dams, was that the Great Depression had ended, the Pacific Northwest had become more industrialized owing to the war, and the population had grown, as large numbers of defense workers had decided to stay on. During the war by far the biggest users of BPA power had been the giant aluminum consumers: Boeing Airplane Company, Kaiser Shipyards, Hanford Engineering Works, and Bremerton Navy Yard, and aluminum producers like ALCOA. The aluminum producers shut down in 1945, but a year later a number of plants reopened; there was a continuing demand for aluminum for national defense in the emerging bipolar confrontation in the world, and there was a buildup of new uses such as aluminum siding. Rural electrification, which had come to a stop during the war, picked up again. Domestic appliances and business machines multiplied. The power boom was on.

Mac was put to work on McNary Dam on the Lower Columbia River, at

Umatilla Rapids. He did his design work in the planning section because there was no other place to do it. While he was at it, the Portland District of the Corps decided to set up a formal design section. We must look again briefly at office politics, for it led Mac to make another move, his final one. The chief of the Bonneville Hydraulic Laboratory, which had been shut down during the war, wanted to be head of the new design section, and because he was important he was given the job. His assistant was a man who had worked with him at the laboratory. They knew about making models of dams but nothing about designing them. Because Mac was the person who was doing the actual design, he was transferred from the little group around his former supervisor with whom he had a good understanding to the new section to work with new people with whom, it turned out, he did not. The chief's assistant became his new supervisor. He had several ranks above Mac and several more years of school, but Mac had to teach him the elements of hydraulic design, an unsatisfactory situation. Mac, who disagreed with him on almost every important point, was considered uncooperative. More important, he was unhappy. He tried to get transferred back to his supervisor's little group, without success. The final straw was his supervisor's pick an assistant, a structural engineer who likewise had no experience in hydraulic design. Mac's career looked to have reached a dead end just as it was getting started. He badly wanted out.

There was to be a way. Three local events bearing on Mac's career as a dam designer and on hydroelectric power production in the Pacific Northwest occurred at this time, in 1948. The first event came about naturally, an excess of a good thing, river flow. Construction at McNary had just got started when the region was hit by a flood on the Columbia River nearly equal to the record flood of 1894. A combination of deep snowpack, heavy rain, and unusually warm spring weather caused flows on the Columbia and Snake Rivers to peak at the same time, a dreaded combination. The reservoirs behind Grand Coulee and Bonneville Dams proved entirely inadequate to prevent serious flooding in three states, resulting in the worst disaster in the river's history. Vanport, a wartime construction town built on the flood plain between Vancouver and Portland, was completely swept away, with considerable loss of property and a number of deaths. From the point of view of McNary Dam and the Columbia Basin development in general, the flood was providential. By drawing attention to the need for multipurpose dams to control the Basin's rivers, it sped up Congressional appropriations. The Columbia River's projection of both menace and promise made an impression in Washington, untying the government's purse strings.

The second local event in 1948 was a revision and updating of the 308 Report, published as House Document 531, 1950. The Corps of Engineers and

27. Columbia River Flood at Vanport, 1948. The largest public housing project in the country, Vanport was thrown up in a hurry in 1943 to accommodate shipyard workers in Portland and Vancouver. It was built on reclaimed lowland, and during the 1948 Columbia River flood, on May 30, a 200-foot section of dike gave way, releasing a ten-foot wall of water, drowning the town, with a loss of at least fifteen lives. Local officials had been lax before the flood, and government officials were after it. The disaster drew attention to the need for multipurpose dams for flood control. US Army Corps of Engineers.

the Bureau of Reclamation were both involved in drafting the plan, their rivalries peaceably quieted in the interest of a coordinated strategy for the region. The plan downplayed irrigation, and because Bonneville Dam was thought to have shown the way to preserve fish runs it did not foresee a major problem there. Because of the flood that year, the plan naturally included flood control, but its main emphasis was on navigation and hydroelectricity. The most important change in the region since the original 308 Report was the enormous growth of power usage, which accompanied the growth in population and industry, themselves brought about in part by the power delivered by the original two federal dams. Over the ten-year period between 1937 and 1946 the demand for power in the Northwest had increased threefold, and it was expected to triple again over the next fifteen years. As if to prove a point, the Pacific Northwest experienced it biggest power shortage that winter, with peak demand exceeding the total output of all the power plants. The plan was aimed

28. Portland, 1948 Flood. The free-standing General Grocery Building, a composition of reflections and symmetries. Mac the amateur photographer took this photograph in Portland during the flood that destroyed the neighboring city of Vanport.

29. Portland, 1948 Flood. This cheerless, overcast photograph of billboards and telephone poles has the feel of popular film-noir movies of the time. Albert L. McCormmach.

30. Portland, 1948 Flood. Men building a dam in downtown Portland out of sandbags. Congress in its Flood Control Act of 1950 hoped to replace temporary dams like this with permanent river dams. Albert L. McCormmach.

at meeting this demand as far as possible and at generating further economic growth. To this end it proposed several multipurpose and storage dams. The director of the Bureau of the Budget said of the report that the works projected by the Bureau and the Corps represented a construction program of about twenty years: "when completed they will make the Columbia River system the greatest source of hydroelectric power in the world and will yield additional benefits in terms of flood control, navigation, reclamation, and other beneficial uses of water We are only at the beginning of a tremendous development program in the Pacific Northwest." For a time after the war, opponents of new dams such as fishers had looked like they might prevail, but in 1948 the prospect for a long season of dam building had never looked more promising.

With the renewal of the 308 Report, we see that power was prominent in the Corps' thinking about the Columbia Basin in 1948, just as it had been when it submitted the original report seventeen years earlier. Power was the most evident benefit of multiple-purpose dams, and it is not difficult to see why. Irrigation and navigation directly affected a relatively small number of people, and flood control paid off only when there was a flood. Power was different: it entered the homes and places of work of all the people, and it flowed all the time. It paid for the dams; more important, it brought wealth to the region, the foundation of the good life and a secure nation. By reason of its versatility, power went far to satisfy most of the desires behind the building of dams. The other uses of dams could not compete with what power gave back.

The third and from Mac's point of view most important local event in 1948 was a transfer of work away from Portland. The Corps of Engineers is organized into geographical divisions and districts, defined by watersheds. Today there are eight divisions in the US and forty-one districts in the US, Europe, and Asia; Portland is the headquarters of the Northern Pacific Division as well as of the Portland District. New districts are only infrequently created, but this happened in 1948. Initial work on McNary Dam was started by the Portland District, which set up a field office near the dam site, but continuing work on McNary and new work on the upcoming dams on the Columbia and Snake Rivers promised to swamp the district and also the other existing Northern Pacific district, in Seattle. To accommodate the increased workload, the Northern Pacific Division set up a new district in southeastern Washington, the Walla Walla District.

An echelon of some eighty employees in the Portland District had been assigned to work on McNary Dam and other funded projects marked for the new district. The chief engineer of the new district came to see Mac and some others in the echelon, to see if they were interested in moving. Some were, while others saw Walla Walla as remote, and they wanted to stay put. There followed a difficult period of divided possibilities and loyalties.

Mac was not put off by the new location. He had grown up in Pendleton after all, a town just across the Oregon border from Walla Walla (and one of the towns that had been considered for the new district). What Walla Walla had going for it was precisely its location. Near the Columbia River and its largest tributary the Snake, it was right in the middle of the Corps' expanding activity. An old frontier town with stately white mansions and high, arching shade trees, Walla Walla – Mac again anticipated correctly – would become a center, if not the center, of the dam-building era in the Pacific Northwest. When in the summer of 1949, the Portland echelon was disbanded and all of its work transferred to Walla Walla, Mac quit his job in Portland and moved to Walla Walla. Two hundred and fifty other Corps employees recruited from all over the nation moved there at about the same time, and this was just the beginning. A year later the Walla Walla District had nearly 600 employees, and two years later nearly 1,100. Employment levels in subsequent years varied greatly depending upon the projects at hand.

Five miles east of Walla Walla, there is a short, steep rise off to the left of the highway, leading to a large airfield built by the Army Air Force during World War II. Part of it is now the regional airport, used by planes that fly between the towns and cities out there. To the side of the runways there was a warren of one-story, wooden buildings, originally a military hospital, which the Corps of

Engineers converted into offices. In these buildings, Mac worked for twenty-five years. During these years, the Corps was a world leader in designing hydroelectric dams, and a good share of its work was done in those converted World War II buildings. Its glory days behind it, the Corps is still there in Walla Walla, only now it is housed in a splendid building of its own in the center of town.

At the beginning, Mac rented a room in a vacated barracks at the airfield. His family – now grown to five – stayed behind in Portland until the house could be sold. On Fridays after work, Mac filled his car with commuting engineers like himself and drove with his foot to the floorboard down the Columbia River Highway to Portland, where he worked all day and most of the night Saturday and most of Sunday turning the attic into a finished room. On Sunday night he drove back up the Columbia River to Walla Walla, at that same high speed. Timid engineers did not ride with him a second time. When the house in Portland sold, the finished attic made no difference to the price. Neighborhood, the real estate agent explained to Mac, counted for everything. Mac's family joined him in a little rented house in a courtyard at the edge of Walla Walla.

As in Portland, in Walla Walla hydraulic design had a small section of its own. Had Mac made up his mind sooner, he might have headed this section, but by the time he made his move, another man had the job. Older than Mac, this man knew about government routine, how to evaluate employees and deal with all sorts of paperwork. He was a capable supervisor. It was just as well that he had the job, Mac decided. As head of the section, Mac would have had a higher rating and salary, but he would not have taken to the supervising end. The way it worked out, when Mac lifted his eyes from his table, they fell on the back of his boss a few feet in front of him, as often as not filling out a form. Mac was left to do what he liked to do, design.

There was, however, a definite downside to Mac's move. With the end of World War II, normal bureaucratic practices had returned. Mac had left Portland well-known only to arrive in Walla Walla unknown and, formally speaking, under-qualified. It took years for the college dropout Mac to reestablish himself in Walla Walla. (Mac had been taking night extension courses in Portland to earn a degree, which he had to give up.) In the meantime, the strain of the job took a toll at home. The family broke up, and Mac took a room in the YMCA.

31. Mac, 1950. He had recently joined the new Walla Walla District of the Corps of Engineers when this photograph was taken. He got better looking as he grew older; in this photograph, he bears a passing resemblance to Paul Newman. Albert L. McCormmach.

LAST FEDERAL DAMS

In the new district, Mac continued with his work on the hydraulic structures of McNary Dam. (We defer the discussion of this dam to Part 4 of this book, which contains photographs.) When construction was completed in 1954, seven years after it had begun, President Eisenhower gave the dedication speech. He said he wanted private enterprise to have a share in the hydroelectric development of America's rivers, but in other respects he sounded much like Theodore and Franklin Roosevelt, making the Progressive-era case for the maximum, multipurpose use of the nation's natural resources. "It is essential that every drop of water, from the moment it falls upon our land, be turned to the service of our people. Thus we will save our soil . . . develop power, prevent floods, improve navigation, and supply our tremendous and growing domestic and industrial needs for water." Dams, Eisenhower said, were vital to the region's "comprehensive planning" for the "full use of the water resources of this entire river system," and in the case of big dams like McNary, which were too expensive for private business, it was right for the federal government to step in. Mac, who turned out for the dedication, would have heard in the president's words an affirmation of his career as a designer of big federal dams. The ceremony was held in the cavernous McNary powerhouse; in light of the priority of power in the development of the Columbia Basin, the setting was appropriate.

There were to be more new starts on federal dams in the Columbia Basin, but the government did not dominate the development as it had up to then. The combination of a Congressional cutback of appropriations for federal dams and tax incentives for private power development resulted in four hydroelectric dams being built by utilities in the Middle Columbia River in the 1950s and 60s. Something similar happened on the Middle Snake River. The Bureau of Reclamation proposed building a high dam in Hells Canyon upstream from the four dams the Corps of Engineers planned on the Lower Snake. Unlike the Corps' dams, which were routinely approved, the Bureau's projects were sometimes turned down by Congress, and this is what happened with the Hells Canyon high dam. Idaho Power Company was licensed to build three low dams on the Middle Snake instead. From the standpoint of the Northwest Power Pool, it made no difference whether the power came from federal or private dams; since 1957, the BPA had had made its transmission lines available to transmit non-federally generated power.

There was a lull in dam-building after McNary Dam, and Mac did military work for a number of years. The Walla Walla District of the Army Corps of Engineers was in charge of all military construction east of the Cascade Mountains in Oregon and Washington and also in Idaho and Montana. There was a good deal of it then, since America had a new enemy in the Soviet Union, and the drums of war were beating again. There were lots of antiaircraft-gun emplacements around Hanford, all highly secret, and there had to be roads to get to them. Mac designed the culverts for all these roads. The Corps built a number of new airfields in the Northwest, which required tanks for storing fuel and pipelines to transport it. Petroleum engineering is not exactly hydraulics, "hydr" stranding for water; strictly speaking, it is fluid mechanics, but it falls under hydraulic design all the same. Petroleum has a different vapor pressure than water; fuels vaporize at a much higher temperature than water, and pipes have to be designed to take this into account. Pipes at airfields had to deliver a certain number of gallons per minute to so many planes simultaneously in a certain length of time. Planes had to be fueled fast, and they had to be emptied fast. Mac was the only person in the district who knew how to do this work, and so all of it came to him.

The John Day Dam fell within the Walla Walla District's territory. The Corps' 308 Report proposed a dam at the John Day Rapids, roughly midway between Bonneville and The Dalles Dams. It was authorized by Congress two years after the 1948 Vanport flood, when the memory was still fresh. The initial plan was for a very high dam capable of impounding a great quantity of water for flood control: no more Vanports. After a study submitted to Congress in 1956 revealed that a reservoir that big would do a lot of damage in the John Day reach of the river, the storage capacity was reduced to a quarter of that originally planned, though even this scaled-down reservoir stretched seventy-seven miles. The final plan was for a low, run-of-the-river dam like the Corps' other Columbia River dams. John Day Dam was no longer viewed as Portland's confident protector but as a "last ditch stand" in the event of another big flood. (Portland can still be flooded, even after all of the dams that have been built. Although there is a large amount of storage capacity in the basin, the Columbia is not controlled to the same degree as other river systems such as the Colorado and the Missouri; the reason is not that the other basins have greater storage capacity – they do not – but that the runoff of the Columbia is so much greater, second only to that of the combined Mississippi-Missouri system.) Mac did a great deal of work on this last federal dam on the Columbia River. Part 4 of the book has a discussion and a group of photographs of the dam.

John Day Dam was meant to produce a lot of power. When it was authorized, the plan had it down for 1,200,000 kilowatts, already more than McNary. Later the plan was revised to take into account a California power interconnection in 1964 and also peak loads – originally thermal plants covered peak loads in the Columbia Basin while dams provided the base load, but this arrangement was reversed in the late 1960s, when hydroelectricity could no longer meet the requirements – increasing the number of generating units to deliver 2,160,000 kilowatts, with skeletal units in the powerhouse for future generators, bumping up the power even further, second only to Grand Coulee's. The extraordinary growth in the demand for Pacific Northwest power in the thirty years between Bonneville and John Day Dams is evident from a comparison of the hydroelectric capacity at John Day and the initial planned capacity at Bonneville of 86,000 kilowatts, which at the time was expected to flood the market for power.

When John Day Dam is added to the dams in Canada, the dams built by utility districts, and the other dams built by the federal government, it makes a total of fourteen dams on the Columbia River, and these are only the mainstream dams. Ice Harbor Dam on the Snake River, Dworshak Dam on the North Fork of the Clearwater River, Lucky Peak Dam on the Boise River, and other dams on the tributaries of the Columbia River add a good many more to the total. It works out that big dams appeared in fairly quick succession, on an average of over one a year. The development of the Columbia Basin was completed in its major features in a period that is short as historical periods go, within the span of one well-timed career. To anyone who gave a thought to the magnitude of the investment and of the alterations of the course of nature, the development was breath-taking . . . or alarming, depending on the person's point of view.

Mac received a reserved-seat ticket for the dedication ceremony at John Day Dam, sponsored by an influential pro-lock-and-dam lobby, Inland Empire Waterways Association. Upon presenting his ticket, he received a name tag with a five-inch, purple ribbon bearing in gold letters "Dedication." The main address was given by Vice President Humphrey, who was at the same time the Democratic presidential candidate in the coming election. He was a New Dealer who liked the Columbia power system, and he was also attuned to currents of thought that were then sweeping the country. In his address he made the best of it. America was on the "threshold of a new era," he said. Multipurpose projects like John Day were fine but they were not enough any longer. "Instead of the uses to which nature can be put, our attention should be directed to the environment." This remark, which might seem subversive of the spirit of the occasion, which was after all the completion of a multipurpose dam, reflected the ambitions of the speaker and the temper of the times. The dedication of the last dam on the Columbia could be taken as a memorial as easily as a celebration. With the turn Humphrey gave to it, its timing marked the passing of the developmental era and the beginning the modern environmental, the earlier receptive to big dams like John Day, the latter rejecting. The building of John Day ended, as it had begun, in politics. The vice president pulled the big switch. It came with the job.

32. John Day Dam Dedication Name Tag, 1968. Today the gold lettering is much faded. Albert L. McCormmach.

33. Award Ceremony, 1969. In this group from the Walla Walla District of the Army Corps of Engineers, Mac is in the second row, third from left. The occasion was a nationwide award, jurors selected from several technical societies; John Day Lock and Dam received "First Place Award for Engineering Design." Albert L. McCormmach.

Humphrey's warning proved prophetic on the occasion, for the dedication ended in a minor environmental disaster. Only one turbine was ready, and so nearly the entire river passed over the spillway, entraining air that was forced into solution, with the result that 20,000 migrating adult salmon and millions of juvenile salmon died of supersaturation (discussed later) at the dam that day.

The start-up of big federal dams in the Columbia Basin came to an end with four Corps of Engineers dams on the Lower Snake River, completed between 1961 and 1975. They had been authorized in 1945, but for ten years their construction was held up for a combination of reasons: partly budgetary, partly fish concerns, and partly because Eisenhower put a temporary hold on new dams. Washington Senator Warren Magnuson and the waterways lobby fought long and hard for the dams, citing reasons of the usual kind, paramount among them inland navigation and hydroelectric power for national defense. In the latter connection the Atomic Energy Commission, they argued, needed the dams to meet the demands of Hanford's reactors for supplying the explosive of choice of the Cold War, plutonium. In 1955, Congress approved the first funds to get things started.

Ice Harbor, the Snake River dam closest to the juncture with the Columbia River, was the first to be built, and the first hydroelectric dam to be approved during Eisenhower's administration. Construction began in 1956, two years before John Day, and it was completed in 1961. The major dedication speech delivered the next year was by another New Dealer Vice President Johnson, who praised Magnuson for promoting the dam and said that he looked forward to all the kilowatts that would be generated when the other three Corps dams on the Snake were ready, all serving to secure democracy and freedom. The desire for more hydroelectric power still seemed boundless. Johnson's confidence in the Columbia Basic multipurpose development contrasted with Humphrey's reservations at the John Day Dam dedication six years later. As at John Day, at Ice Harbor on that celebratory day a disturbing event dampened the spirit, with a hint of the prophetic. After the speechmaking, two sky divers jumped from the dam, one falling in the water and the

34. Ice Harbor Dam. This 100-foot-high, concrete gravity dam is the first of four similar Corps of Engineers dams on the Lower Snake River. Construction on the dam finished the year this downstream aerial photograph was taken, 1961. In the foreground is the lock, its lift-gate towers visible on the far right. On the other side of the spillway is the powerhouse; to the original three generators, three more would be added fifteen years later, making a total capacity of 603,000 kilowatts. There is a fish ladder on each side of the dam; the one on the far shore is just visible in this picture, and the one on the near shore is hidden by the lock. No water is coming over the spillway; this photograph was taken for the record, not to show off the dam. US Army Corps of Engineers.

other landing on a pile of structural steel breaking a leg. From the point of view of their advocates, Ice Harbor and the other Snake River dams made bad landings too, in the courts.

Lower Monumental and Little Goose, the next two Snake River dams, were completed around the same time as John Day. Preliminary construction on the last dam, Lower Granite, began in 1965. In 1969, the year after Humphrey spoke of the importance of environmental considerations at John Day Dam, sport and conservation groups and Idaho's senators called for a ten-year moratorium on dam-building on the Snake, which included shutting down work on Lower Granite. The proposal died, and the next year the Association of Northwest Steelheaders, a sport-fishers group, together with other similar groups brought a suit against the Corps of Engineers to stop construction on the dam.

That year, 1970, was the year the environmental movement in America came of age. Twenty million Americans attended coast-to-coast rallies, in what was to be was the first of an annual event in protest against the degradation of the environment, Earth Day. That same year, the most important environmental legislation to date, the National Environmental Policy Act, was signed into law. Public support was fast moving away from building new big dams, and the half-built Lower Granite Dam was caught in the middle.

The Corps filed a motion to dismiss the suit about Lower Granite and proceeded to begin main construction on the dam. The district court ruled that only Congress had the power to stop construction, whereupon litigation was moved to an appeals court; the appeals court sent the motion back to the district court for reconsideration. Lower Granite was completed in 1975 (or in 1979 when the power plant was completed). In 1977 the district court ruled in favor of the Corps of Engineers, determining that the issue was moot since this dam along with the other three dams on the Snake River were up and running. Litigation concerning the management of migratory fish at all the Snake River dams has continued to this day. What has not continued is the startup of any more big federal dams in the Pacific Northwest.

Mac remembered that when he started out as an engineer in the 1940s, people at the Corps talked about "conservation." Conservation meant using natural resources effectively, without waste or destruction. Government at all levels believed in conservation, the Sierra Club believed in it, Mac believed in it, it was the right way for America. It was an inspiration. By the time John Day Dam was finished, conservation had acquired a wider meaning and inspiration. At the dedication, Humphrey said that "multipurpose" projects like John Day, instruments of the earlier meaning of conservation, overlooked an important purpose: henceforward, the goal should be "all purpose conservation," which takes into account not just human needs but the "total environment," quoted above. People, however, often meant opposing things by the two words. The environmental movement was, in part, a grass-roots protest against the conservation goal of maximum, efficient use of America's natural resources. Lower Granite might extract the last kilowatt from a river, but at an unacceptable cost to the environment. The political and legal activism of environmentalists was directed against run-away industrial development and large-scale technology, judged responsible for pollution and loss of habitat and natural beauty. Environmentalists were on an eventual collision course with federal big-dam builders; in league with fisheries they would prove a formidable political force.

PLIGHT OF THE FISH

"Run of the River" is an engineering term for relatively low dams designed to pass rather than store water, dams intended primarily for generating electric power and improving navigation. With the exception of Grand Coulee Dam, run-of-the-river is the type of dam the government built on the Columbia and Snake Rivers. The term unintentionally characterizes the problems dams pose to migratory fish. In the state of nature fish are accustomed to having the run of the river. Now that we manage the river, the fish depend on us; they have what rights we decide. The engineers' term can be put as a leading question, Which party do we wish to allow first rights to the run of the river, humans or fish, or can we speak of equal rights?

Migrating fish on the Columbia River are legendary. The Chinook, or king, salmon is the best-known of the several species of Columbia salmon, each of which is made up of a number of breeding stocks. The somewhat smaller steelhead is well-known too. (The Columbia River steelhead is usually called an ocean-going trout, but it belongs to the same genus as the salmon.) Born in the fresh-water tributaries of the Columbia and Snake Rivers, juvenile salmon and steelhead go down the river to the ocean, where they spend several years, growing to very considerable size, a mature Chinook salmon weighing twenty pounds or more. Near the end of their lives, while they are still in excellent physical condition, they start the journey in reverse, their destination the tributary near the very place where they were born; their sense of smell tells them when they get there. They might journey as far as 900 miles, all uphill. They do not eat along the way, so that by the time they arrive they are spent and ragged. Their last act is to spawn, after which they drift downstream and soon die. This was their hard life before any dam was built there.

Through the peak period of dam-building, the 1950s and 60s, problems of the salmon seemed manageable. Columbia fish runs seemed to be holding their own. The annual catch of salmon remained at around ninety percent of what it had been over the half century before any dams were put up, from 1880 to 1930, though hatchery salmon had much to do with this. The reassurance was premature; it takes observations over many years to judge reliably the health of salmon runs, which can vary from one year to the next by a factor of ten. The seeming disaster that the dams turned out to be for the fish only became clear in the 1970s, and it only became a major public issue from the beginning of the 1990s after it was found that some of the breeding stocks were likely to die out; the Snake River coho salmon had become extinct, and other Snake River runs were identified as threatened and endangered.

The high mortality of migrating fish in the Columbia Basin has many causes, which are no mystery. They include ocean warming; over-fishing in the rivers and the ocean; loss of spawning habitat owing to logging, grazing, farming, mining, urban development, and other human practices; diminished river flow owing to water withdrawals; pollution by chemicals from fertilizers, pesticides, and pulp mills, and the most visible cause of all, the eight federal dams that a fish born in Idaho must pass to reach the ocean to grow up and then pass again on its return from the ocean to spawn.

Long before any dams were built, Columbia salmon were on the decline. In the nineteenth century this was caused by commercial fishing and especially canning, and to a lesser extent also by sport and Indian fishing; in the early twentieth century it was caused more by loss of habitat. The Corps reported on the decline to Congress as early as 1887. Fifty years later, in its 308 Report, the Corps predicted that new dams would cause a further decline. When it came time came to build its first dam on the Columbia, the Corps incorporated fishways in the design from the beginning.

When Bonneville Dam was built, the main worry was the passage of salmon migrating upstream, and the main recourse was fish ladders, which were known to work. The very earliest fishways were fish ladders; records going back three hundred years show that branches were bundled together to create stepwise pools in deep stream channels to guide fish around obstructions. Although the Corps had little previous experience, the pool-and-weir fish ladders it installed at Bonneville appeared to work fine at first, and people believed that they were going to do a lot of good, as they did, only not enough good.

Locks and elevators for lifting fish began in the mid-1920s, when dams got to be higher than about fifty feet, and the Corps incorporated these in its dams too, though they were found to be less effective than ladders. It was assumed that the juvenile salmon migrating downstream would probably make it through the turbines unharmed; all the same Bonneville Dam was fitted with bypasses to divert them from the turbines. (These were structures resembling fish lad-

35. Salmon Wheel on the Columbia River. Mac's father took this photograph in 1912 while traveling on a stern-wheel steamboat. Once a common sight on the Columbia River, salmon wheels could harvest a prodigious number of fish. Commercial fishing contributed to the early decline of salmon runs, and salmon wheels along with traps and seines were banned before any dams were built on the river. J. W. McCormmach.

36. Fish Ladder. This is the Oregon shore fish ladder at Bonneville Dam; the dam was built with three fish ladders. The water moving through the channel is a small river in its own right. The author.

ders, only they were narrower and their steps higher.) Given the reasons for building dams and the little understanding of fish runs of the 1930s, it is noteworthy that as much attention was given to fish as there was.

Following the authorization of McNary Dam and the Snake Rivers dams after World War II, government fishery agencies came around to accepting them, if reluctantly. Government biologists were unhappy, but they went along with the dams all the same. In the face of the nation's need for dams, they knew that a decision had been made, and that their concerns carried little weight. There was a decided hint of fatalism in what they had to say on the matter. The US Fish and Wildlife Service acknowledged that the economic importance of the rivers justified their multipurpose development, and that they could not be considered "solely for the sake of maintaining salmon runs." The interior department, which in 1947 had proposed a ten-year moratorium on building dams in the Columbia Basin to study their effect on fish, acknowledged that the economic development of the rivers was of such national importance that the "present salmon run must, if necessary, be sacrificed." The Oregon and Washington state fishery agencies acknowledged that multipurpose dams were essential to the "advance of civilization," and they would try to preserve fisheries only so long as it was "economically possible." The agencies did what they could to see that in the construction of dams, consideration was given to fish runs, but they would not hold back the hands of the clock.

A Corps biologist in the 1940s characterized the attitude of some of his engineering colleagues, "To hell with the fish, I'm here to build a dam." How widespread this attitude was is hard to say, and no doubt there were engineers who thought that way then, but it was not Corps policy. The Corps gave consideration to fish, as required by the Water Power Act of 1920; the trouble was that biologists did not know much about the habits of fish, and designers of dams did not know just how little they knew about the effects of dams on fish.

The same year that the proposal was made to build more dams, 1948, the US Fish and Wildlife Service reported a fifteen percent overall loss of fish at Bonneville, an order of loss expected to be compounded with each new dam. In 1956, the Oregon Fish Commission determined that salmon migrating upstream were delayed several days at Bonneville by the ladders and by the slack water behind the dam, and these fasting fish have only so much energy to make it to their spawning grounds. Long-term studies of the physiology and behavior of the migrating salmon were begun then, and research on fish facilities picked up. It was accepted that under the best of circumstances, the dams would take a toll on the fish runs, and in 1957 the Corps began its Fisheries Engineering Research Program in an effort to mitigate the harm caused by the new dams. In 1958 Congress amended the Fish and Wildlife Coordination Act from

1934 – this act had required the Corps to consult with fish and wildlife agencies on the operation of dams – to require the Corps to do more, to give "equal consideration" to fish and wildlife as to water development and to compensate for any losses as fully as possible. The hope was that engineers and biologists working together could improve fish facilities.

Fish ladders help spawning salmon, but they do nothing for the other half of the salmon migration, the downstream juveniles. These little fish go through the turbines, over the spillways, or through special bypasses. Unlikely as it appears from the violence at the foot of the spillway, if the dam is the low, run-of-the-river type, most of the juveniles survive that waterfall and the subsequent back-flow eddies of the stilling basin. The turbines tell a different story. Contrary to the original thinking based on a report by the Corps in 1934, a substantial percentage of the juveniles passing through them are killed; this corrected finding entered the 1948 revision of the 308 Report. The juveniles are killed not by the whirling blades of the turbines, again as was first thought, but by the sudden changes in pressure. If they are not killed in passing the dam, whether they go over the spillway or through the turbines, the little salmon come out downstream under conditions of shock and disorientation. These stunned survivors become easy prey for seagulls, squawfish, and other predators. Juveniles, like adults, are delayed in their journey at each of the dams, exposing them to a variety of risks. Variations in the pool elevation, the velocity of flow, and the temperature of reservoirs all affect the health and mortality of juvenile and adult salmon in complicated ways.

The juvenile salmon faced an additional problem. The National Marine Fisheries Service determined in 1971 that three quarters of the juveniles passing the dams on the Lower Snake River were destroyed by supersaturation, by gas bubbles in the blood. This was the result of the absorption of air by water passing over the spillways. When water moves with a velocity of around twenty feet per second, it begins to entrain air; when it moves with the much higher spillway velocities, it entrains more air, which mixes with the water. The air that water coming off a dam absorbs would soon escape in turbulent water, but below a dam there is no turbulence to speak of, since the pool behind the next downstream dam backs up to it. Instead, under pressure the air goes into solution in the river and enters the bloodstream of the fish, damaging or killing them, a condition known as bends in humans. Control of river flow by upstream storage projects can eliminate the need for spilling water, an obvious solution to the problem. There is another solution, which Mac was first to suggest. Structures called "flip lips," or less colorfully "spillway deflectors," are installed low on the face of the spillway to deflect the water coming off it, causing turbulence; the deflected water is dispersed over a large area, limiting its depth of plunge and pressure. Flip lips work best under low flow, and they are especially helpful where there are a series of dams, since supersaturation is

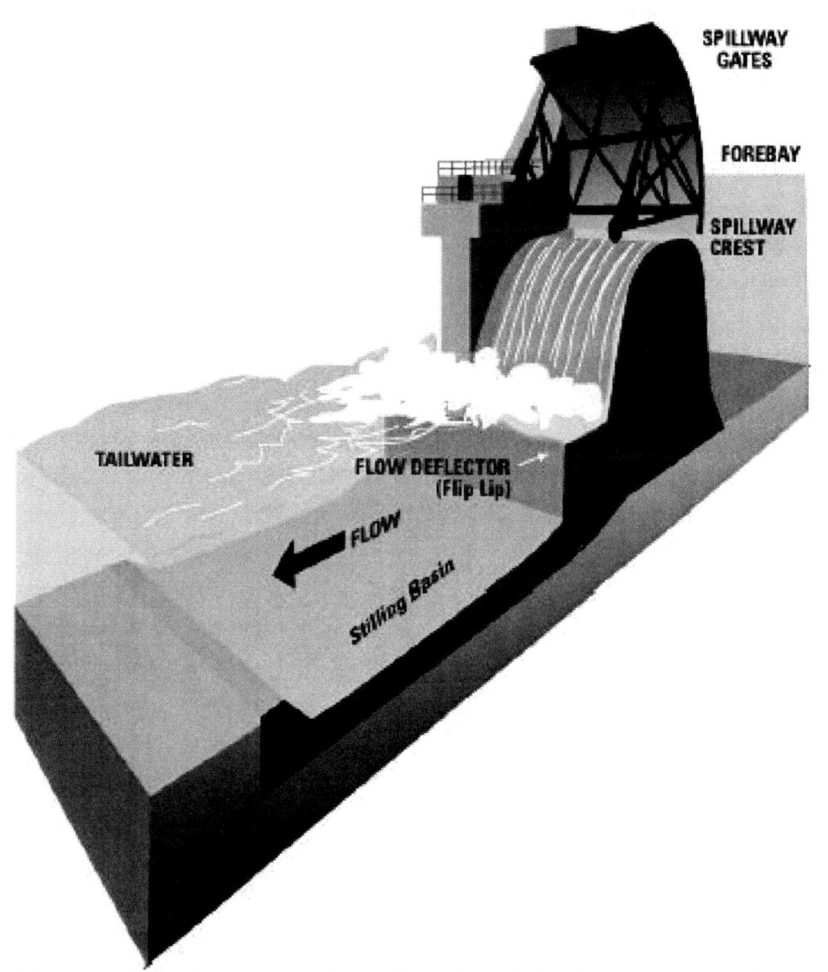

37. Flip Lip, Drawing. This shows a flip lip in action, deflecting the flow at the base of a spillway. US Army Corps of Engineers.

cumulative. They are installed on most dams on the Columbia and Snake Rivers.

Bypass systems of various sorts – submersible traveling screens, deflectors, and openings – directing juvenile fish away from the turbines have helped, as have screens for preventing them from getting trapped in irrigation canals. A bounty on squawfish probably has not helped, since there are so many of these predators. Fish are raised in hatcheries and released to go to the sea, but these fish prey on, crowd out, and reduce the genetic diversity of wild fish, and they return in uncertain numbers too. Dams affect peak flows and river depth, which in turn affect the sandbars and beaches the salmon use as spawning grounds.

Artificial spawning channels have been constructed above McNary Dam and elsewhere in the hope that fish born in them will return, with mixed results. When the severity of the supersaturation problem was first recognized, the National Marine Fisheries Service under contract with the Corps began to transport juvenile fish past the dams, using every means available, trucks, barges, and airplanes. This method spared the fish having to go over the crests of dams or through the turbines, and until the 1990s most fish were transported and released below Bonneville Dam, but now this is done only for fish at three dams on the Lower Snake; and fishery agencies now allow only fall fish to be transported. Rivers are monitored for dissolved gases and temperature, and the dams are operated to try to maintain healthy levels. Turbines have been designed to cause less harm to the juvenile fish caught up in them.

The latest improvement is – like flip lips, screens, and bypasses – a structural change. It actually raises the height of the spillway. This might seem a poor idea if it were not that when juvenile fish move down the river they are no more than ten to twenty feet beneath the surface, whereas the crest of the spillway can be fifty feet beneath the surface. When the dam is spilling water, the fish have to dive way down to find the opening, where the pressure and velocity are high, and this is stressful on them and it delays their passage. Engineers and biologists designed a "removable spillway weir," which can be raised during fish runs, effectively raising the height of the crest to the level at which the fish are moving. The weir is a steel structure attached to the upstream face of the dam by hinges, which allow the weir to be rotated and submerged at times of flood flows. Removable weirs have been installed on the federal dams on the Snake River.

In 1980 Congress established the Northwest Power Planning Council, a state-appointed body which in planning for future regional electric power is required to consider environmental costs and energy alternatives, and the BPA is required to finance any measures it decides on to help fish and wildlife. The council promptly recommended drawing down the reservoirs of dams to speed up the flow of the river with the object of flushing the juvenile salmon downstream faster. This was the first proposal to change the operation of the dams, and it has since been adopted, though at a loss of revenue from power and at a cost to irrigation, navigation, aesthetics, equipment, and even the stream bed. Beginning in the 1990s, in connection with the required Environmental Impact Statements, the big dam-builders of the Pacific Northwest, the Corps of Engineers and the Bureau of Reclamation, and the BPA regularly conduct what is called a Columbia River System Operation Review. This has multiple objectives, but as is to be expected the issue of migratory fish has come to dominate the operation of dams.

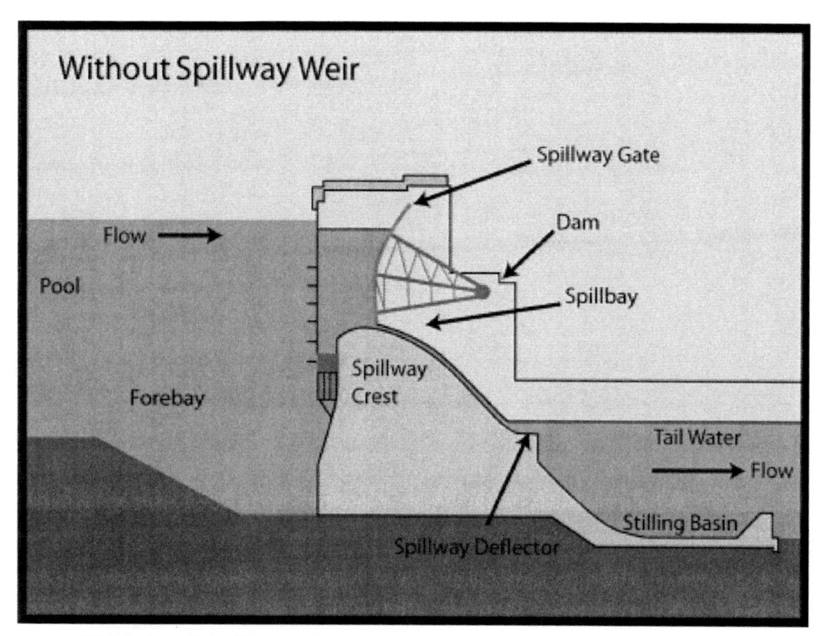

38. Normal Spillway, Drawing. US Army Corps of Engineers.

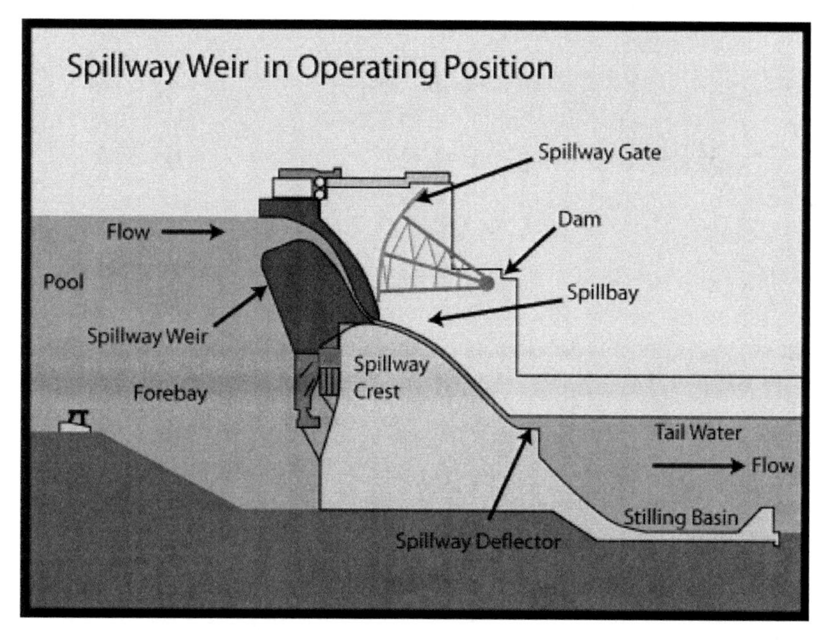

39. Spillway with Movable Weir, Drawing. By raising the height of the spillway, the movable weir enables more juvenile salmon to pass the dam safely. This inventive machine has been installed at Snake River dams on a trial basis, as an alternative to the possible removal of the dams. During high flows when the dam needs to spill more water, the movable weir rotates from a hinge at its bottom, coming to rest on the pad shown at the far left, out of the way. US Army Corps of Engineers.

We see that many partial solutions to the problem of moving fish safely past the dams have been tried; ingenuity has not been in short supply. As it now stands, people are not prepared to give up power and the other benefits of dams, and they are not prepared to give up fish, especially the salmon. Unlike many threatened or endangered species such as the Spotted Owl in old-growth forests in Oregon and Washington or five snail species on the Middle Snake River, the Columbia salmon are conspicuous for their size, numbers, and confinement to the river. Everyone knows what salmon look like and where to look for them. Everyone likes salmon. They are a commercial resource, they are highly valued by sports fishermen, and for many preservationists, particularly in the Northwest, they stand for nature whole. When they do not come back, there is a big hole in nature.

40. Chinook Salmon. National Park Service.

The Corps of Engineers, which built and which now operates most of the Columbia Basin dams, has taken most of the heat for the plight of the salmon: called fish killer and public enemy of the environment, it is accused of overreaching, of excessive pride, of seeking to control nature at whatever cost, and its dams are cited as evidence. The Corps's sister dam-building agency the Bureau of Reclamation is considered a nature wrecker too and on the same grounds. Those persons who place the highest value on the natural world see a greedy intention behind every move the two agencies make. Whatever merit the charges have, they tend to assume that the Corps and the Bureau are of one mind, and that it is always the same. The evidence does not support these assumptions.

In response to the environmental movement, the Corps began to change in the 1960s from an agency mainly concerned with building structures to an agency concerned with managing natural resources. In that decade, it employed seventy-five specialists on the subject, and in the next decade, 500. In 1970, it employed 200 biologists as well. The Walla Walla District, which was considered "pretty progressive in terms of the environment," together with other districts set up environmental sections. In the last half of his career, Mac worked as much with biologists as he did with engineers.

It appears that, by law and by choice, the Corps and the Bureau are as responsive to the needs and opportunities of the environmental era as they were to those of the New Deal. (They still support "conservation," emphasizing its contemporary meaning, broadened from the earlier sustainable use of resources to include the preservation of biodiversity.) Likewise, by and large, activists who oppose river development in the name of the environment do not want to stop the world from turning. Many people would agree that we need institutions with a long perspective to restore and to sustain depleted migratory fish stocks. The reinvented Corps and Bureau of the environmental era are candidates for the needed institutions, though many activists and critics remain deeply skeptical, convinced that big bureaucracies cannot truly change their ways.

For all of its environmental problems, the Pacific Northwest is still a natural wonderland compared with other regions of America, and with the world at large. Relatively speaking, even the salmon have it good here. The intensive development of the Columbia River Basin was repeated in every major river basin in the country, but nowhere else has so much effort and money been expended on fish. Washington and Oregon (to a greater extent than their neighboring states) still have some healthy salmon runs, and they have key watersheds and some pristine waters. Precisely because there is so much to preserve, environmentalists in the region feel a particular sense of urgency, and dams top the list of their concerns. (What is often forgotten in the heat of the controversy is that the same environmental groups that target dams also target a variety of large-scale structures and widespread practices in forestry, mining, waste disposal, and so on. Dams are part of a much broader debate on the development of natural resources.)

Dams are a proud achievement of the conservation era. Its sequel, the environmental era, has some proud achievements too and several working laws to its credit, in particular, the Endangered Species Act; since its passage in 1974, the law has been on the side of fish. It is an unfinished era, and the stakes are high. If its representatives avoid the temptation to overlook humans in their concern with the environment, this too may be remembered as a remarkable era in American history.

For several decades, Northwesterners have looked for a human ladder to get them past an impasse: an agreement between adversaries that will hold up over time, an acceptable compromise that will leave the economy sound and that will assure the survival of salmon runs. There is no assurance they have found it yet. There is no sign either that people are about to give up.

ENDINGS, AN ERA AND A CAREER

When I was in college, Mac sent me a greeting card designed by the Army Corps of Engineers for its employees to send out at Christmas time. The front shows the logo of the Corps, a red, twin-tower castle. Inside there is a picture of power generators at McNary Dam, which had just gone on line. The good tiding reads, "More power to you!" Mac circled the message in pencil, "Pretty clever, eh?" I kept the card, not realizing that one day it would capture a bygone era, when power was king, or messiah... I had heard people speak of the electrical millennium. I since learned that Roosevelt repeated the old saying about more power to you with reference to the Columbia, and maybe the person who designed the card knew this too. In any case, the designer assumed that people would be glad to know of a new powerhouse, that its picture was good public relations for the Corps. The Corps moved with the currents of history then, and it still does. Today it likely would not make up a Christmas card showing electric generators. A king salmon? Probably not either, since it might suggest complacency.

Mac's career fell in an extraordinary period in US history, second in significance only to the Civil War. His period was preoccupied with the economic, political, and military strength of the nation. The Pacific Northwest contributed its share. Federal dams of the Columbia Basin, as a part of Roosevelt's New Deal, offered relief during the Great Depression, and at the same time they showed Americans that their political and economic system still worked. More than that, they showed the world what America at its best is about. In his speech at the dedication of Bonneville Dam in 1937, Roosevelt said that great public works like this were proof that unlike dictatorships abroad, America directed its wealth to improving life and increasing the nation's prosperity, not to amassing armaments for war. (The dams would be caught up soon enough, but the buildup of armaments was not an American initiative.) The New Deal is often described as a federal effort to save capitalism in a time of its seeming collapse. From this perspective, the first federal dams on the Columbia were not an alternative to free enterprise; on the contrary, they were a commitment to its endurance.

Federal dams did more for the economy of the region than did any other development in the twentieth century. They improved river transport, controlled floods, irrigated land, earned revenue, attracted industry, and provided power. They vindicated Roosevelt's investment of public money and political capital in the region. He had had a good idea about electricity, to make it available at low cost to people everywhere as well as to industry, and he had had another good idea about a way to bring it about quickly in his day, to build big dams with hydroelectric capacity using federal funds. His dams empowered the Pacific Northwest.

The big-dam era in the Columbia Basin came to an end, as it must. Its main accomplishment was evident at the time: it had made the Columbia Basin the greatest producer of hydroelectric power in the world. Late new starts, as we have seen, were the Corps of Engineers dams on the Snake and Kootenai Rivers, and if we wished we could end our account with the year the last Snake River dam was completed, Lower Granite in 1975 (or in 1979). Or we could end it with the construction of the second powerhouse at Bonneville Dam in 1974-82 and the third powerhouse at Grand Coulee Dam in 1967-74, the latter acquiring its final generator only in 1980. The years 1980-82 being the last of any significant expansion of big dams, we could say that the first two big federal dams in the region were also the last dams of the big-dam era in the Pacific Northwest. The year 1980, we note, saw the passage of the Northwest Power Act, with its environmental constraints on power development. The Pacific Northwest is in step with the rest of the country: very few large dams have been built in the US since the early 1980s. In the middle of the twentieth century, when Mac set out on an engineering career, the right combination of fields for the Pacific Northwest was power and hydraulics. Dams held the promise of the future. Since the late twentieth century, the right combination has been power and environment. This is as true for the country as it is for the region; President Obama appointed a White House aide to coordinate energy and environmental policy. We live with a different future than Mac's generation.

Over the course of Mac's career in the Corps of Engineers, the Pacific Northwest was the proper place for him. In other parts of the country, the Corps was busy dredging, clearing and snagging, maintaining and restoring, and in Walla Walla it did some routine flood-control work and the like, but its main work there was designing, building, and operating new structures. Toward the end of Mac's time, all the big projects were finished. He stayed on for a few years after he reached retirement age, but the excitement had gone out of the work for him. His last job was about protecting a little town in Oregon from high water, and it was a very boring job. (It was very boring but not insignificant. The town was Heppner, which in 1903 had been hit by a flash flood caused by a cloudburst and collapse of a debris dam, killing about 225 of its inhabitants.) Mac realized it was time to quit.

That meant a little party and a little ceremony. A photograph records it. It shows the colonel in uniform, shaking Mac's hand and holding a certificate in his other hand. What is interesting here is what Mac is wearing, a black leather jacket. In his later years at the Corps, he had taken to riding his motorcycle instead of driving his car to work. That was unusual at the time. On the day of the ceremony, he rode his motorcycle; too late, he realized that he probably had not dressed appropriately for the occasion. His outfit suggests that he managed to spare a moment to receive the honor and that he was in a hurry to get back on his motorcycle, as he no doubt was.

Mac's colleagues took him to a favorite tavern of the Corps, where they gave him a pool cue that unscrews in the middle and comes with a carrying case, very professional. They also gave him a framed sketch done by one of the more artistic engineers. It shows Mac in uniform, though it is not an army uniform, which he did not have, but the outlaw uniform of the modern West. He resembles a Hell's Angel, looking very determined. He is astride a hog – Mac's actual motorcycle was good-sized, but less than a hog – front wheel in the air, making puffs of smoke indicating terrific speed. The cartoon is an affectionate joke with an edge, and it is not bad, capturing a side of Mac. The Corps saw him off with a hustler's cue and a picture of him as a hellion-on-wheels. Some of his colleagues may have had mixed feelings, but they had lost a good engineer, and they knew that.

During his working years, Mac's appraisal of his coworkers was often critical, but with distance he was generous. In retirement, he said that it took a lot of smart people to build those dams. I have wondered what he seemed like to those smart people. Well, he was determined, the first thing people noticed. He was smart too, and knowledgeable and impatient, probably the next things. He thought the world had no scarcity of fools. He was unforgiving of carelessness, particularly under a cloak of authority. Because he was thorough to a fault, he was rarely wrong, and he did not care who knew it.

One day at the Corps' cafeteria, where Mac and a number of other engineers were eating lunch, one of them produced a beautiful red apple from his sack, which he polished on his sleeve and laid in the center of the table. There were murmurs of appreciation. He named the apple, it was a such and such. Mac picked up the apple, turned it over, studied it, then laid it back and said, "It's not a such and such, it's a Jonathan." The engineer took his apple back. It had come from his tree, he knew what it was. Mac said, "I will bet you a thousand dollars it is a Jonathan." The table fell silent. Mac finished his lunch. When I asked Mac how he could be so certain , he gave me an exact description of the characteristic knobby formation on the bottom of that variety of apple. The engineer with the apple did not speak to Mac for a year. This was the kind of trouble Mac got into by being a know-it-all. The variety of apple, what did it matter? But being right about a dam did matter. That was one of the reasons he was an engineer. You had to take Mac as he came, all of the parts, and if you did that you came out ahead.

Mac spoke his mind freely, and he spoke it very well. He had at his command, maddeningly, always the precise word, and he had a sharp wit. He also had very considerable charm, but how much of that side he showed to his colleagues is in question. They gave him plenty of room, that much is certain. When you see Mac coming, one of his colleagues said, you better get out of the way or he will run you down. If they found him difficult, they also found him dependable and capable, a man of integrity, whom they wanted on their side.

As the oldest of Mac's children, I can say that for those many years when he worked on the dams of the Columbia Basin, he was a largely absent father.

41. Mac in Retirement. He is wearing a tie and smiling a little, both rare in a photograph. The author.

He had no social life, and neither did his family. For all intents and purposes, his work was his life. Everything else took second place. His work was – as indeed any work can be – at times a battlefield, but it was also, and most important, the one activity through which he could realize his better nature, his penchant for excellence and public service.

With Mac as much as anyone I have studied as a historian I recognize the interplay of historical forces and individual choice. The forces were the Great Depression, World War II, postwar prosperity, and the Cold War. The choice was an engineering career in federal power development on Pacific Northwest rivers. If power is given its comprehensive meaning, Mac's life can be seen as a confluence of human and physical power lines. When I was growing up, I was too close to Mac and dams; I could feel the lines of power through Mac, but it would be a long time before I understood them.

Thanks to advances in science, medicine, and technology, human lives become longer, and historical periods shorter. Mac had a long, full career through the developmental era, and he lived almost as long into the environmental era. When he retired, he told me that given a free choice of power or fish, he would have chosen fish. He would have preferred never to have put a dam on the river, and not only because of the fish; a natural river has a beauty that a stagnant pool behind the dam does not. But dams were going to be built, no question about it, and Mac thought that if there were going to be dams, they should be built well, as they were. He did not think that it was wrong to build the dams at the time. On the contrary, at the time it was the right thing to do; they improved lives, and they helped a nation at war. Rather he was sad that despite good fishways, and despite the well intentioned management of the fish, the salmon of the Columbia Basin seemed to him headed for extinction.

From a professional standpoint, Mac was proud of the work he did as an engineer. In his specialty, he believed that he and his colleagues were ten years ahead of the textbooks, that in effect they wrote the textbooks. With every new dam came a new set of problems never before solved. They were building something both technically demanding and important to society. But Mac was not proud of the consequences, the interference with the free flow of the rivers of the Columbia Basin. He came to believe that fish and dams are incompatible, that there is probably no in-between, and that an honest choice has to be made, fish or dams. Experience strongly suggests, he thought, that unless the dams are taken out, the outlook for migratory fish in the Columbia River is poor. Mac, who designed some of their beautiful hydraulics, would have liked to see them blown up and every last chunk of concrete removed from the river. For practical reasons, politicians are unlikely to take that course, and given the impact of other activities on the river, the removal of the dams probably would not do much for the fish in any case. But we are talking about Mac's desires here.

What might seem a cruel twist of fate did not seem that way to him. He could contemplate with equanimity, as environmentally desirable, the removal of engineering structures he had spent his working life designing. His passionate engagement with his work was a thing of the past, but he had never felt more strongly about the fate of the earth. The window on the front door of his house in Walla Walla was plastered with decals of environmental groups, to which he made regular contributions.

In my account of the power development of the Columbia River, I em-

42. Environmental Stickers on Mac's Front Door. The author.

phasize the play of historical forces and overlook the contribution of individuals. This is because the account is brief, not because I think individuals are unimportant. Strong personalities in national and local politics, in lobbies, in the Corps of Engineers and Bureau of Reclamation, and in the environmental movement all left their mark on the development. Ordinary individuals like Mac, whose names are unfamiliar, also belong in the account, somewhere. They might have strong personalities, Mac did, but they were in no position to shape events. Their power was of a different kind; it derived from the organization they served, to which they contributed their specialized knowledge and skills. What these ordinary individuals working with a common purpose brought into the world is remarkable.

We have lived with the development of the Columbia River Basin for so long it no longer looks remarkable to us, and we have to remind ourselves. We tend to take the dams and power lines for granted, as we do interstate highways and international air travel, as part of the scene, as part of the routine of our ordinary lives. Technology has moved on and with it our attention, which is presented with perpetually new objects of wonder.

What can still seem remarkable to us are people from another era, who can be as near to us as our parents, who brought giant power to the Northwest. Mac's story may suggest as much to some readers. By remarkable, I do not have in mind any remarkable ability of Mac's. Mac once said to me he thought he had a talent for engineering, which he probably had, but then so did his colleagues. Nor do I mean Mac's insistence on getting things right, which he brought to his work as a hydraulic designer of dams. That is what every responsible engineer brings to his work. What is remarkable, anyway what seems remarkable to me, is Mac's way of looking at large human endeavors in their setting in the natural world, this together with his feeling for both, the endeavors and the natural world. The word I give this quality of Mac's is sympathy; it is from this quality that his career derives its poignancy. By Mac's sympathy I mean his feeling both for what was admirable in the dam-building era and for what is admirable in its successor, the environmental era. He fully recognized the difference in the values of the two eras, and he was clear about which of them he supported, but this in no way conflicted with his sympathy, which was encompassing. When I give the quality the word sympathy, an everyday word after all, it does not sound remarkable, and maybe it is not. Lots of conservationists became, or always were, environmentalists, I realize. Yet in my experience I rarely come across a sympathy of Mac's kind.

His sympathy came in part from his idea of the way society changes direction. He saw the hydroelectric power development of the Columbia River Basin as an example of projects that society undertakes from time to time with the goal of bringing about major changes. Large numbers of people come together, they overcome resistance to the changes, and they see the projects through to the finish. What they accomplish is seen to do a lot of good, and that is what people think about. Then after a time, it is seen to cause a lot of harm, and then people think about that. The good is still there, and now there is the bad. The clarity and simplicity of the project, which seemed to fulfill so many desires at the beginning, are gone, replaced by uncertainties. Hydroelectric dams gave people power, which changed their lives for the good, and they changed rivers in ways that down the road proved bad for fish, which changed people's thinking about dams. People settled on a new set of desirable changes and went in search of new projects suited to the environmental era.

Mac's sympathy had other sources too. One was his logical mind, capable of taking in new information and drawing useful conclusions from it, with which he might correct an earlier opinion of his. He had a thorough grasp of the issue of dams and fish, the evidence, the arguments of the various sides, the technical and political problems, the complexity of this particular reality, the developed rivers of the Pacific Northwest. A related source of Mac's sympathy was his objectivity, a scientific habit of mind, mistakenly assumed to be contrary to sympathy. He did not feel a personal stake in the way the world used to be run, even if he had helped make it run that way. He was quite free of that kind of vanity and that kind of nostalgia. He was perfectly open to well-informed and well-argued environmental objections to the old way. For many years, he worked to try to save the fish runs; in the end, he doubted that it could be done

A final source of Mac's sympathy was his understanding of the two sides of nature, the physical and the living. Chance encounters, as we have seen, directed Mac toward an engineering career; if they had been different, he could willingly have gone to work in a fishery and wildlife agency or an agricultural agency, and it would have suited him just as well as engineering. He had an extensive knowledge of wild plants and animals, based largely on his own observation, from which I got my education in the natural world while I was growing up. Throughout his life, he was equally drawn to living nature and to the physical forces of the natural world. When he was a boy he read with profit several heavy-going volumes of *Luther Burbank, His Method and Discoveries,* which his father had accepted from a car buyer in default on his payments. Mac converted the large vacant lot next to his house into a productive vegetable garden for his family and an experimental plant-breeding laboratory. At the same time he was fascinated by electricity, including radios, which were still new; to the neglect of his homework, he stayed up nights exchanging information with fellow avid amateur radio operators about building receivers and transmitters. In college he took classes in agriculture and engineering, in more or less equal numbers. During World War II, while he was starting out as an engineer, he turned a vacant lot on his street into a Victory Garden from which he supplied the neighborhood with fresh produce. In his work as a hydraulic designer of dams, he collaborated on a regular basis with fish biologists. When he retired, gardening was a major part of his life. His observations as a naturalist and his long study of both biology and the physical sciences and engineering made him receptive to the opportunities of both eras, the earlier of big dams and conservation, the later of environmental protections.

In retirement, Mac was the local library's most faithful patron. He looked over all the new acquisitions, checking out any having to do with the sciences. He told me that he was grateful for the sciences because they made life so interesting. Alternative chance encounters early in life would not, I am certain, have led Mac to take up any of the arts, literature, painting and sculpture, or music. He responded strongly to natural beauty, and he had a skill in photography and design, so he had an appreciation. The English Romantic John Keats, who is much admired for his insight into the mind of a poet, thought that a person of achievement, especially a poet, has an ability to accept the irresolvable nature of some things, a cultivated open-mindedness. He called it "negative capability," which he described this way: it is "when a man is capable of being in uncertainties, Mysteries, doubts without any irritable reaching after fact & reason." On the central issue of Mac's later work as a dam engineer, the fate of migratory fish, the first half of Keat's statement applies; Mac was in uncertainty for a long time, as were many of his colleagues. The second half of Keat's statement does not apply to Mac, who as an engineer worked to resolve the issue, and fact and reason were his tools. He did not have a poet's option and or a poet's capability in Keat's sense. To be drawn to the arts, a person does not have to fit Keat's description, but I never saw Mac take an interest in any of them.

Keats wrote his own epitaph, "Here lies one whose name was writ in water." These words do not apply to this poet who wrote it while brooding about the malice of his enemies – everybody knows his name – but it does apply to most people on earth who have lived and died, Mac for example. In this connection I think of the title James Agee and Walker Evans gave their classic book on Southern sharecroppers' lives in the Great Depression, *Let Us Now Praise Famous Men*, taken from a passage in the Book of Ecclesiastes, "Let us now praise famous men, and our fathers that begat us," which continues: there are men that are "renowned for their power . . . that have left a name behind them, that their praises might be reported. And some there be, which have no memorial; who are perished, as though they had never been," and these men, who are probably our fathers, deserve our remembrance and our praise too. Mac is not as obscure as a sharecropper, and it is not as if he had never lived. Indeed, the Columbia Basin is full of his monuments, rivals to the tombs of ancient kings, the big dams, but his part in it is all but anonymous, like that of the ancient tomb masons. This book on hydroelectric power is a token remembrance, which can be taken as praise of a man who in the ways of the world is without renown.

When I was young, I did not distinguish between Mac and dams. Mac and dams seemed about the same size. When I grew up, I could separate the two, though the separation was always incomplete. He still seems large, but for other reasons, among them his sympathy. Mac, who distrusted compliments, would, I trust, have accepted this estimation.

PART 2: DOWNSTREAM JOURNEY, MAC LATE IN LIFE

After he retired from the Corps of Engineers, Mac only once seriously considered returning to his profession, and then only temporarily. Around the time that the damming of the Columbia River Basin got going in earnest, the 1950s, other countries around the world had similar ideas for their rivers, and Pakistan was dreaming irrigation and power. Pakistan had rivers and a scheme for developing them, the Indus Valley Settlement Plan. The Indus River and its five tributaries, the Western Rivers, drain the western Himalayas, converging in the middle of Pakistan to meander across the plain to the Arabian Sea. Shared by Pakistan and India, the Indus Basin already had a number of impressive dams designed to control floods and generate power such as India's 741-foot tall Bhakra Dam. A key development in the Pakistan plan is Tarbela Dam, which stands 485 feet from the riverbed and is two miles wide, known as the biggest earthfill and rockfill dam in the world. Its hydroelectric power capacity is large, equal to half that of Grand Coulee Dam, three and a half million kilowatts, and its storage capacity is truly enormous, over eleven million acre-feet, surpassing Grand Coulee's. If its benefits are impressive, so are the problems that have plagued it ever since its reservoir was filled. The tunnels that were used to fill the reservoir failed, and the reservoir had to be drained to avert a disaster. The reservoir was found to be full of sinkholes, which had to be covered, and downstream a huge plunge pool formed, threatening the main spillway, and it was the same with an auxiliary spillway. The Indus carries a great deal of silt, which builds up behind the dam. Environmental damage is extensive. Constant maintenance is required to prevent a tidal wave from sweeping down the densely populated valley. The New York firm that Pakistan hired to design Tarbela approached Mac and some other experts on big dams, offering them handsome consulting fees to go to Pakistan to work on the problems. In the end, Mac declined the offer, in part because he was unenthusiastic about helping dam up another great river running through another beautiful valley.

When Mac retired, he bought a portable tape recorder, which he tried out on trip he took from Walla Walla to Portland on family business. He recorded casual recollections and impressions as he drove down the river he had worked to master (a word he never used) during his career. So far as I know, it is the only tape recording he made. What follows is in part a paraphrase of what he said and in part my elaboration.

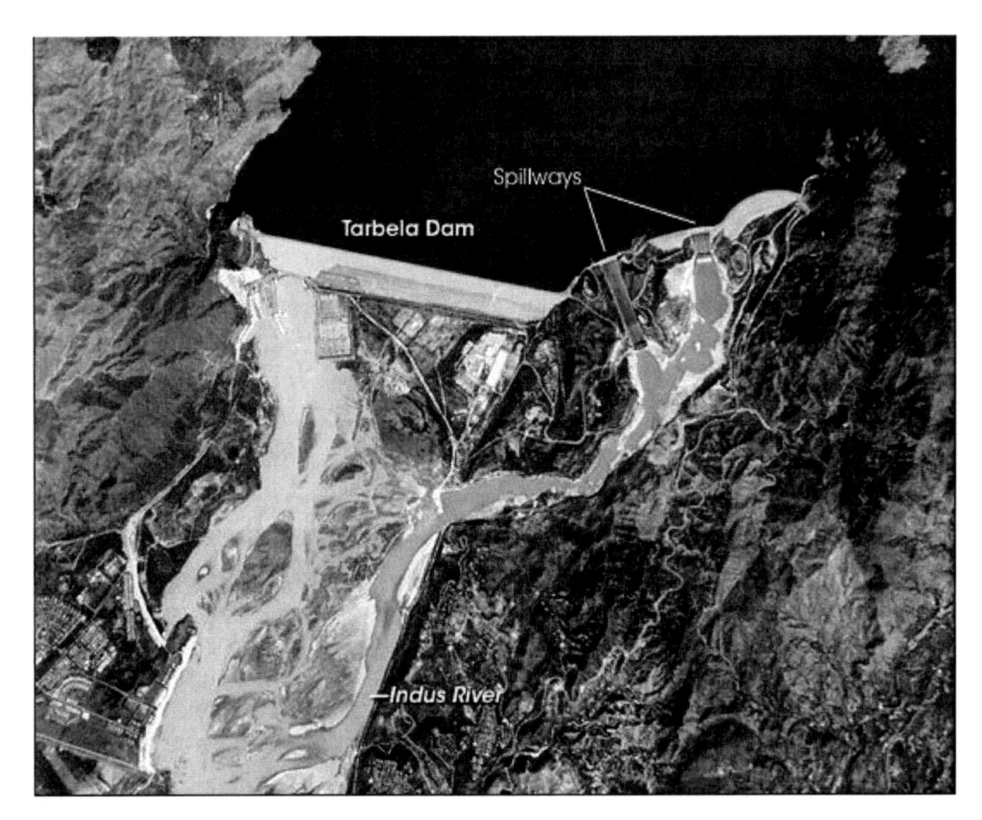

43. Tarbela Dam. Pakistan's Tarbela Dam on the Indus River, the largest earthfill dam in the world, photographed from space. It has a number of similarities with Columbia Basin dams. Around the time that the US and Canada signed a treaty on the Columbia Basin, in 1960 Pakistan and India signed the Basin Water Treaty, which resulted in the Indus Basin Project. Tarbela Dam was built in 1968-76 as part of that project. Like Columbia River dams, it is a multipurpose dam; its reservoir serves Pakistan's irrigation system and provides limited flood protection, and its generating facilities provide a good share of Pakistan's hydroelectric power. It likewise has environmental problems including a reduction in numbers of certain species of fish. NASA.

On that July afternoon, the outdoor thermometer on the porch of his little, turn-of-the-century wood house next to a railway spur read a mild 85°. Mac climbed into the small Japanese car he now drove. On a pad of paper, he noted the odometer reading, his custom when beginning any trip, 8576.9 miles. He dictated this fact into his tape recorder, which he then laid on the empty passenger seat beside him.

Mac headed west from Walla Walla on a two-lane highway past irrigated fields of alfalfa, barley, and wheat. Harvest was under way, and little self-propelled combines were moving on the hillsides. Soon the Walla Walla River came into view as the highway began to climb, the valley narrowing to between a hundred and fifty yards to a quarter-mile across. Mac was pleased that his

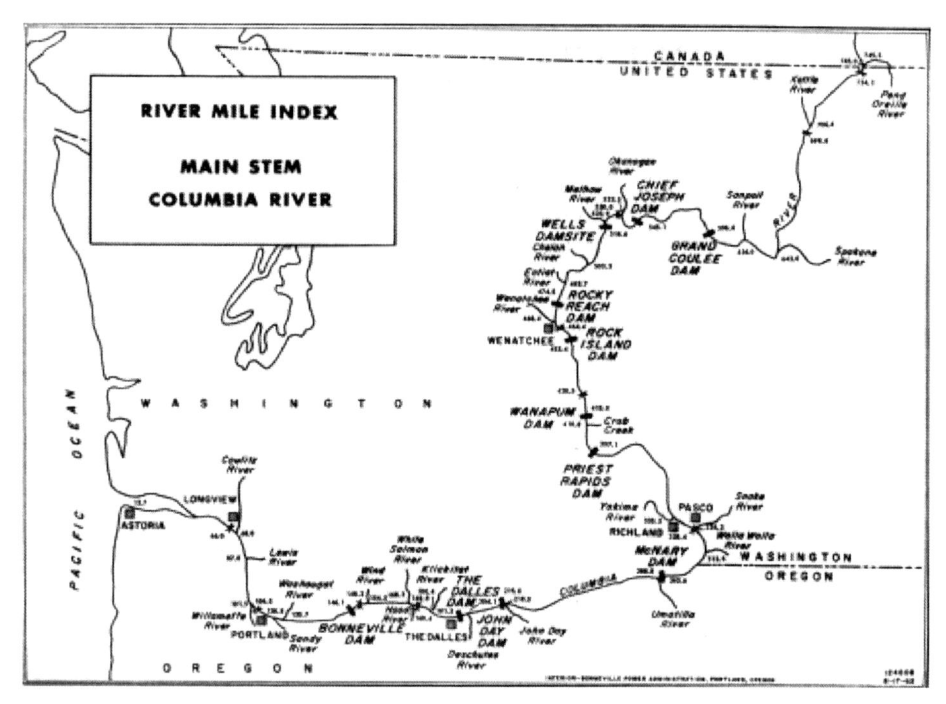

44. Mileage Map of the Columbia River. The river mileages agree with Mac's odometer readings on the interstate, the highway hugging the river. From McNary Dam to the last dam, Bonneville, is 146 miles by the river, and it is the same number of miles by the highway. Bonneville Power Administration.

little Japanese puddle-jumper was doing seventy in fifth gear. He bought this marvel of efficiency recently, after the country almost ran out of gas. That time the shortage was artificial, an embargo, but one of these times it would be the real thing. The dams were probably there to stay. People were going to want all the power they could get.

The irrigated valley was replaced by dry hills and sagebrush. At Pasco Junction, at 8606.5, to the right, the Snake River meets up with the Columbia River, and a short way up the Snake is its first dam, Ice Harbor. Mac did the hydraulic design for it. The initial main impetus to put Ice Harbor and subsequent dams on the Lower Snake was navigation, and for a while after the dams were finished navigation on the river increased, but then it declined. That was not the case with power, which only increased. Ice Harbor is half the size of Columbia River dams, with half the power, but since there are three more dams upstream from it, each putting out more power than Ice Harbor, it adds up. Idaho wanted Ice Harbor for navigation and irrigation, but Congress put up money for it primarily because of its hydroelectric power.

Just beyond the junction to the Snake River lies Tri-Cities – Pasco, Kennewick, and Richland, the atomic city – and the Hanford nuclear facilities. The Corps of Engineers had been in there at its nuclear beginning, to prepare the site, surveying and emptying a tract thirty miles across. Broken up by volcanic slabs and ditches run dry, it had a forbidding look, so Mac thought. What struck him on the many trips he made there for the Corps were the jackrabbits. There were so many of them, and they were so hungry they ate the surveyors' wooden stakes down to a point. People who lived around Hanford wanted to shoot every last one of those jackrabbits, and they tried. What decided the fate of that hot, dusty place was its location on the Columbia River, with its ample water for cooling the nuclear reactors and its bottomless pit of hydroelectric power to run them. High-voltage lines were strung from Bonneville and Grand Coulee Dams to a substation at the site.

"And there it is!" Mac having taken the highway to the left at Pasco Junction, he came to the Columbia River just below its confluence with the Walla Walla River. There were the rugged, prismatic canyon walls of Wallula Gap, the biggest of the water gaps through the basalt topography of this part of the Columbia Basin.

The water there was just slightly rippled, and aquamarine in hue. Before the river had been backed up behind McNary Dam, Wallula Gap had been much deeper, but it was still impressive. Cliffs alternated with steep, rounded hills, some with rim-rocks, some without, all brown and sear, tending toward yellow at this time, but beautiful nonetheless. Tapered slopes at the foot of the cliffs were partially covered with sage and grass and in places were bare. Mac wanted to explore the cliffs with someone.

Four or five hundred feet above the highway, a succession of lava flows came into view: one, two, three, four, five, each having a slightly different formation, and each with slightly different colors to the rocks. There was some bending of the lava flows in the gap, but they were mostly horizontal around there, indicating that there had not been much uplift, in keeping with the absence of mountains.

45. Wallula Gap. The author.

46. Near Wallula Gap. The author.

McNary Dam! At the time Mac moved to Walla Walla, in one little dry area the first-step cofferdam at the McNary site could just be seen, the first bit of construction. In Portland, Mac had already determined the shape of the spillway. Later, when the dam was finished and he saw water come over his spillway, he experienced a deep satisfaction. The flow agreed with his calculations and drawings perfectly. Mac's first dam, like first love, held a place apart in his affections.

A mile or so past the dam, Mac pulled into Umatilla, the odometer reading 8634.5 miles. This little river town had to be moved above the reservoir created by the next downstream dam, John Day. It had some life when McNary was under construction and workers were looking for someplace to go and something to do, but to a passing motorist like Mac, it was just a sleepy, very long strip of motels and cafés.

Mac came to some fertile fields with movable irrigation rigs, interspersed with rolling, sandy country covered with sagebrush. The contrast between the

47. Columbia River Bluffs. Succession of lava flows. The author.

48. McNary Dam. This run-of-the-river, concrete gravity dam at the Umatilla Rapids was the third federal dam on the Columbia River. The Corps of Engineers' 308 Report called for developing sites lying between the bookend sites, Bonneville and Grand Coulee; McNary is roughly halfway between the two. Its hydroelectric power capacity is 980,000 kilowatts, about the same as Bonneville's. In the photograph, the dam is shown spilling water, as viewed from the Oregon side. This dam is discussed in detail in the text. The author.

two lands, irrigated and dry, side by side, was striking. How a little water can change things! Mac was struck, though he had seen it countless times. People who made their living around there were thankful for water.

The non-skid road surface near the dam was noisy, and Mac could not tell what gear he was driving in by the sound. He saw that it was fourth rather than fifth, defeating the purpose; he shifted up. Construction of McNary Dam began at the same time as the horsepower race. Mac remembered that Cadillac was first, coming out with a big V-8 engine in 1949. In the 50s, Detroit made lots of two-ton, 300-horsepower gas-guzzlers, fifteen miles to the gallon, ten miles, nobody kept count because nobody cared about efficiency then, gas was practically free. Cars got bigger still in the 60s, 400-horsepower, say. Hydroelectric power was the same story, the people wanting more and more. The Corps of Engineers more than obliged them, building big powerhouses up and down the Columbia. This was the time of Americans' love affairs with big cars and big dams, with big power really. Both affairs came to an end of sorts in the 70s, and for the same reason, costs, but gas was no longer a giveaway either, and the costs were of a different kind. Of course, Detroit went on making big cars, but big dams did not come back.

Beyond Umatilla, the two-lane highway from Walla Walla turned onto the east-west, four-lane interstate highway, I-84. Mac read 8649.8 at the junction. Just beyond it was the Dodge City Motel, where Mac invariably stopped for a coffee break on his hundreds of trips along the river with the Corps of Engineers. This time he did not stop for coffee, since he was not feeling the least tired, and besides he had brought his own. He was now driving into the sun, and because he had neglected to set the rear window in the ventilating position, his small, poorly insulated car was hot. He took a refreshing drink of coffee from a wine bottle.

At Arlington, the odometer read 8680.4. When John Day Dam was built, the reservoir rose above the level of this town too. It was anticipated, of course, and everything worth saving had been moved uphill in advance. A straggle of houses up the sides of a draw was about all that was visible of this town from the highway.

Mac saw some wheat being moved down the river, a common sight. A tugboat was pushing two barges, as usual, but a single tug could push five if it had to. This is because from the standpoint of the tug, the river is completely level between dams, and at each dam the tug with its load is lifted or lowered with no effort on its part, and relatively little expenditure of power is needed to do this. The dams had been good to navigation. There was no controversy about that, at least

At five minutes after seven, Mac came to milepost 125, the distance from downtown Portland, his destination. The odometer read 8694.4. The water had turned gray-green, with white caps. He was driving into shadows.

A cumulus cloud, its whiteness blending with grays and blues, lay on the crest of the bare hills across the river. Downstream two or three miles, the hills were indistinct owing to a haze. The canyon deepened. The occasional scruffy tree along the bank was blown low. There on the highway, 500 feet above the river, Mac's little car bucked a stiff headwind.

Mac would have liked to have a sailboat with an auxiliary diesel engine the right size to cruise the Columbia River. The force of the wind prompted this thought. It was a longing, nothing more. To handle a sailboat, he needed a partner, which he did not have.

Now it came into view. From the interstate bridge crossing the John Day River, Mac could see in the distance the enormous twin towers of the navigation lock of John Day Dam. The sight of these towers distressed Mac, as always. Back at McNary Dam, the gates on the lock were miter gates, which swung open and shut on hinges, like barn doors. The plan was simple and sensible. Then they got the idea of lifting the gates instead of swinging them, and it was a problem from the start. The gates are made of various beams, some of which are perforated, and every time you lift the gates, all kinds of water comes down on you. The head of the Corps of Engineers in the Pacific Northwest was invited to go through the lock at Ice Harbor Dam when it was finished. That was the second mistake. They had put the new kind of gate on the lock there, and the chief got drenched, of course. He ordered that no more gates of that design were to be built. The structural engineers got busy adding drains to the existing lift gates, coming up with what they called "drip-dry gates." Mac supposed this was a sample of engineers' humor. Drip-dry shirts were just becoming popular then. These gates had welding problems, and trash and ice accumulated. They weighed a million and a half pounds, and to lift them required enormous wheels with sprockets for winding cable connected to massive counter-weights. The wheels worked imperfectly, and maintenance was always a problem. When they built the last two dams on the Snake River, they went back to miter gates, having learned their lesson. Mac's supervisor the navigation man was enthusiastic about lift gates. Mac thought they were decidedly inferior to the old gates, a flaw in otherwise majestic locks, which could lift and lower vessels 100 feet, but he had nothing to do with the gates.

At the axis of John Day Dam, the odometer read 8706.5. The spillway, all twenty bays, was dry. This was good news for power, but Mac – now a tourist – was disappointed. Water coming off the spillway was what lent visual excitement to a dam, which otherwise was a largely submerged obstacle to the free flow of the river. Because no water was coming over, Mac was unable to see his flip lips in action at the base of the spillway, but he could see the fish ladders he

49. John Day Dam. Lined up in order: powerhouse, spillway, fish ladder, and navigation lock with lift-gate tower. Because of the perspective, the powerhouse takes over the photograph. If not that dominant in the actual structure, the powerhouse is indeed big, extending over half the width of the river, and in terms of function, the distortion holds a truth. When the dam was built, expectations for the growth of power demand in the region were running high; its hydroelectric power capacity is accordingly high, 2,160,000 kilowatts. This again is a run-of-the-river, concrete gravity dam. Details of the dam are given in the text. The author.

worked on.

He approached Biggs, a collection of gas stations and eateries high up on a rocky bluff overlooking the river, at the intersection of the east-west interstate and the north-south, old Sherman Highway. Long haulers stopped here, as did other folks tired from their travels across the open spaces of the Far West. But Mac was still not tired, and so he did not pull in. The wind on the exposed interstate here was even stronger, in places piling up waves two to three feet high, though in other places the waves were a mere chop. Normally, the towering, snow-capped volcano Mount Hood was in splendid view, but because of the haze, it was completely invisible, to Mac another momentary disappointment.

Beyond Biggs, signs warned of falling rocks. This was crumbly rim-rock country. The rocky spires there suggested to Mac photographic possibilities, and he wanted to explore them. High up there, on a shelf, he wanted to build a little house for himself, wind and all. The sight of this river often set him to wondering about possibilities. He knew it had that effect on others too. It had had that effect on Roosevelt.

Across the river, on the Washington side, the hills were etched by vertical shadows in the draws. The sunlit parts of the hills glowed gold and red, where before they had been gray. The color of the hills was reflected in the river, flecked with white.

50. Power Lines and Lava. A common mix in the Columbia Basin. Albert L. McCormmach.

He was moving across a rolling, high plateau, through which the river had carved a deep valley in its run to the sea. The land dropped steeply to the river, and the hills were deeply eroded and terraced by layers of black basalt, exposed by weathering. The rhythms of this scene had been imprinted on Mac's memory since he was a boy.

The interstate crossed the Deschutes River, the next big tributary of the Columbia after the John Day River. A sand blow came into view. It was irregular in appearance, but at an earlier time it had been rippled, making a beautiful pattern of frozen waves. Mac had taken photographs of it then.

A sign read "Wind Gusts," but Mac did not need reminding. The vertical rock face ahead was familiar to him as the windiest point on the river. Before

51. Columbia River Hills. The author.

the interstate was built, the old Columbia River Highway skirted the base of it, and when Mac would round it, he would be lifted up and blown off the road. Not blown clear off, but that was the feeling. For power, the wind could not compete with the moving water below him, but it could be formidable there. One day people might be needy enough to go after the wind there too, and they will say of the wind what they said about the river, that all this flow should not go to waste. This time when Mac came around the point on the interstate, he did not experience the blast he expected. The trees at the edge of the river did not show violent motion, so there must have been a rare lull in the ferocity of the wind there. That of course was the problem with wind power, its unpredictability.

The Dalles Dam came up, the last dam before Bonneville. The highway is almost at river level there, maybe twenty feet above. The Corps of Engineers built this dam too, but Mac did not work on it, since it was assigned to the Portland District. Just as at John Day Dam, there had been a big rapids at The Dalles Dam. The Corps had a history at these rapids, having built an eight-mile canal with five locks early in the century. Before that, steamboats had

52. The Dalles Dam. This aerial view shows the unique layout at The Dallas. The river is too narrow here for the powerhouse to be aligned with the spillway. Instead it is at right angles to the spillway, parallel to the bank. Built in 1952-57, The Dallas Dam was the Corps' biggest multipurpose dam to that time. Like the Corps' other dams on the Columbia, it is a run-of-the-river, concrete gravity dam designed to produce a large quantity of hydroelectric power, 1,808,000 kilowatts. The dam is also important for river transport, replacing the Corps' outmoded The Dalles-Celilo Canal, dating from 1915. U.S. Army Corps of Engineers.

had to transfer their passengers and cargo to a train, which had transported them to another steamboat line at the other end of the rapids. The same thing happened at Cascade Rapids a ways downstream from The Dalles, where earlier the Corps had also built a canal and locks. When passage at The Dalles opened up, there were celebrations up and down the river. People believed there would be year-round river traffic clear to Idaho, but this did not happen because the river fell too low at times. Later the dams made it possible.

At the head of the rapids there was a spectacular sight, Celilo Falls. When the river was low, it dropped as much as twenty feet at the falls, straight down, with a roar and a huge spray. There were lots of migrating fish in the river, and they had to go up this narrow channel. You saw them leaping by the dozen below the falls. The Indians had built wood scaffoldings out over the falls, which they stood on to catch salmon with dip nets at the end of long handles. The scaffoldings were not much to look at, and the Indians put ropes around their waist to save them in case they fell off. People came from all over to watch them snag salmon in mid air during their runs up the river. They were skilled at it; Mac had often watched them. When The Dalles Dam was finished, the reservoir buried the falls, rapids, canal and locks, everything that had marked this stretch of the river. Indians had been fishing at the falls for hundreds of years. They were given some money, but nothing could make up for a way of life for them, now gone forever. Everyone knew it was going to happen, but the dam was built anyway. It was sad, but given the thinking of that time, there was a solid case for building the dam, which Mac had accepted. Like their salmon, the Indians were a secondary consideration. On these two subjects, the Corps' record was mixed.

There were powerful interests behind the dam, but the dam was not pork barrel. The Corps gets criticized for its wheeling and dealing, and probably a lot of the criticism is deserved, Mac thought. But when The Dalles Dam and John Day Dam went through, it was understood that the Pacific Northwest had a real need for their benefits, especially their power, and power trumped everything else in those days.

The Dalles, now the town, was coming up. Between Walla Walla and Portland, the one town of any size is The Dalles, but even it has only a few thousand people. This was the end of the Oregon Trail. Below the rapids, the weary pioneers piled their wagons on rafts and barges and floated down the Columbia to where the Willamette runs into it. At The Dalles' city center, Mac's odometer read 8733.5. It was twelve minutes to eight.

Just past The Dalles, Mac entered the Columbia River Gorge; a highway sign made it official. Here the parched open land began to give way to a landscape of pine and maple and oak trees. Further on, near the Cascade Mountain divide, this in turn would give way to a humid landscape of densely clustered firs, hemlocks, and more maples and oaks.

The next stop was Hood River, a little, steep-sloped orchard town; the odometer read 8755.5. Just beyond and to the right of the highway was Charburger, where Mac always came to lunch on field trips to the Hydraulics Laboratory at Bonneville Dam. Next to it was the old Meredith Hotel, where he often stayed overnight on these trips. The milepost told him he was sixty-one miles from his destination, Portland.

53. Columbia River Gorge. Wikmedia Foundation.

54. Columbia River Gorge. This view of the gorge shows Vista House, built in 1916-18 as a rest station and viewpoint for motorists on the new (now historic) Columbia River Highway. It stands 733 feet above the river. Wikmedia Foundation.

Mac came to Cascade Locks, a little town named after the canal and locks at Cascade Rapids. It was a ghostly name now; but for a remnant, the locks had gone under when the pool behind Bonneville Dam was raised, Bonneville Lock taking its place. There had been a town before Cascade Locks, Mac remembered it. Whiskey Flats had been a boom construction town when the Corps came in the first time to build Cascade Locks, back in the 1890s. It took its name from the only business in town, saloons. When the job was done, the place quieted down and changed its name. People always seemed to find something to keep them there.

Downstream Mac could see the old cantilever steel bridge over the Columbia River. An enormous landslide had completely blocked the river there a few hundred years ago, and the Indians remembered it in their legends. Bridge of the Gods was the Indian name for the Bonneville Slide, and the white-man's bridge at the site was named after it, made of steel in place of rocks.

The sun dropped behind a cloud bank. Ahead lay the steep, rocky hills characteristic of the skyline around Bonneville Dam, the only federal dam on the Columbia built in mountainous country. The sun reappeared between peaks, looking very red through the mist above the river. Mac could look directly into the sun without difficulty. The color of the sky was bright yellow-orange.

A few miles ahead lay a grayness, seeming to float on the water. The rocks in this stretch of the Cascade Rapids were over-topped by the white man's artificial rock, concrete.

"Bonneville Dam!" This was the last dam, and this was the first, and it was where it all began for Mac. The dam looked weathered, Mac thought, but there could be no mistake about it. It was the work of man.

Everything man-made is evil, what is natural is good. There were some persons who almost believed it, as though humans and tools were unnatural. Those persons were not environmentalists but fanatics, Mac thought. For all the evil to go away, all the people would have to go away. If that happened, in time Bonneville Dam would topple into the river, like a piece of mountain. The river would make a little adjustment, and things would go on. With Bonneville down and the other dams down, the fish would have the run of the river back. The evil would be gone, but nobody would be around to notice. The natural river is magnificent, and so are the dams, in different ways. Drivers can turn off at the viewpoints and take in the scene at their leisure, and wonder at what nature with humans included can do.

Mac could see the two gantry cranes over the Bonneville spillway before he could see any other details. He could identify this dam by the cranes that lift and lower the vertical gates, since most Corps of Engineers dams on the

55. Bonneville Dam. The Oregon shore fish ladder is in the foreground. On top of the dam we see one of two traveling gantry cranes used to operate the spillway gates. Begun in 1933 and completed five years later, this run-of-the-river, concrete gravity dam was the first federal dam on the river to go into operation, the historical beginning of the technological transformation of the region. Details about this dam are given in the text. Albert L. McCormmach.

Columbia and Snake Rivers have rotary gates, which do not need cranes. Every dam is individual in some way.

The politicians were glad to have Bonneville Dam, but they wanted to save the gorge from polluting industries and quick-buck developers. The beauty of the gorge owed to the beauty of electricity, which is that it does not have to be used where it is produced, but can be transmitted over long distances. They knew that the success of Bonneville Dam depended on attracting industries to use the power, and they also knew that these industries could be built somewhere else besides next to the dam. Mac was grateful to them for saving the gorge and for bringing wealth to the region, both. At the time, people really did do the wise thing.

Mac's watch read quarter to nine when he passed Bonneville Dam, the odometer read 8779.3. He looked at the looping high-voltage lines coming off the Bonneville powerhouse. On shore they were suspended from horizontal arms attached to huge steel towers, each a nest of triangles, stacked one on top of the other, the whole balanced on spreadeagled legs. Trees had been leveled to provide these power lines with an unobstructed course and access, a kind of giant's causeway running up the side of the canyon. Mac remembered the BPA's first lines, 230,000-volt circuits between Bonneville Dam and Vancouver and between Vancouver and Portland General Electric. The lines were designed to carry these very high voltages because distances in the West were so great. Bonneville Dam was completed before Mac worked on dams, but as a BPA draftsman he was close. He worked on the routes the power lines took from the dam, and when he became a hydraulic designer for the Corps of Engineers, he studied the dam to learn how it was designed.

Mac thanked his lucky stars that he had left that small-time title business in Pendleton to go to work for the Bonneville Project in Portland. He had sensed that Bonneville was just the beginning, that he would be connected

56. Bonneville Power. The author.

with something that was not going to go away. He accepted the world as it came, and it could be a bitch at times, but it could also be a candy store. There were opportunities for the taking, which you could see if you pressed your nose to the window. In Mac's time electric power was the treat in the window. People wanted electricity. It was going to give them jobs, a decent life. Bonneville belonged to a hopeful future, which was where Mac wanted to be. That was his opportunity. It was up to him. Depending on how smart he was, and how lucky, he would make it or fail.

There were the Bonneville fish ladders, a confidence trick in their way. The dams closed off the whole river but for little channels, and we had a way to make the fish think that these channels were the river and to go up them without losing much time. We fooled the fish, but we fooled ourselves too, Mac knew. We thought we had the answer. We had fish ladders, which were part of the answer, but the other part of the answer was the hard part, and we may never have it, Mac thought. When Mac started out, people believed that Bonneville and all those other dams were a wonderful thing. Fine for the country, we'll have all that power. They did not know it was going to be so hard on migrating fish. If there was a flaw in the hopeful future Mac saw for himself, it came with the dams, the damage they do to fish.

Towering pillars of basalt flanked the highway, giving the Columbia River Gorge a fantastic feel. The roaring of the motor and the slapping of the tires of his sturdy little car reassured Mac that he was still on firm earth. At one time, there was just the river, no mountains, no gorge. Then the land began to lift, and as fast as it did, the river dug its channel that much deeper. Moving water is determined, Mac knew. His work was about determination too, or to use the name we gave to it, about power.

When the land rose, it lifted the streams that emptied into the Columbia. Here they formed a dizzying, vertical landscape of hanging little rivers, the waterfalls for which the gorge was famous.

If it were not so late in the day, Mac would have turned off the interstate onto a restored segment of the original Columbia River Highway. It is a scenic outing for unhurried motorists today, the reason the highway was built in the first place. This first major highway in the Northwest was completed just a few years before Mac got his first driver's license, and it was the first real highway he drove on. He still admired it. It was blasted into the cliff faces of the gorge using the latest construction methods.

57. Waterfall in the Columbia River Gorge. J. W. McCormmach.

This picturesque route, with its stone and masonry balustrades, its snake-back loops, and long tunnels was showing wear by the time Bonneville Dam was built, and it was outdated. People were more interested in making time than in looking at nature. Bonneville Dam forced a major realignment of the highway, which was the beginning of the end of it. The state highway department began abandoning sections of it. Later, pieces of it were returned to their 1920s look as a historic monument to an early engineering triumph. The interstate highway that replaced it is scenic too, but it is down next to the river where it ought to be rather than up high where most of the waterfalls are. You can drive the length of the gorge without slowing down. Like power lines, the interstate cannot afford to zig and zag.

The old Columbia River Highway was a technical feat of a high order, and

58. Old Columbia River Highway. Starting at Astoria, Oregon, on the coast, the highway stretched three hundred miles to Pendleton in eastern Oregon. The section through the Columbia Gorge shown here was considered an engineering wonder of the world when it was built, around the time of World War I. Mac took this picture in 1940.

at the same time it preserved the natural beauty of the gorge, adapting technology to the landscape. Mac remembered a group of French engineers who came over to see what they were doing on the Columbia. They admired the technology of their dams, but they wondered how Americans could build anything so ugly. Mac thought that a functional design can have an appeal of its own, but the French engineers had a point. On a Corps trip to Vicksburg, after the hydraulic laboratory at Bonneville moved there, he made a side trip to see some of the things the TVA had built. He saw Norris Dam, the first proper TVA dam, which by comparison with the Columbia River giants did not put out much power, but Mac thought the structure had more visual appeal, and some famous architects agreed. It was built in the modernist style, architecture from engineering, beauty from function, and so on. A lot of people evidently did not like it for that reason, finding it too plain. They liked another TVA dam better, Wilson Dam, which took its ideas from Greek and Roman buildings, a strange choice, Mac thought, for a facility with the latest twentieth-century technology. Maybe it made the technology seem less strange to the people living around there. Where that left Columbia River dams Mac

59. New Columbia River Highway. This is the Western part of the northern east-west interstate highway, I-84. It begins in Portland, and it roughly follows the Oregon Trail. Through the Columbia Gorge as far as The Dalles, it parallels the old Columbia River Highway. The author.

was unsure. They were plain, no question; you could make of that what you will. The landscaping of the dams was something else again. The TVA did a lot with it, and so did the Corps and the Bureau of Reclamation. They went to trouble and expense to show off their dams and the lakes behind them with beautiful grounds and vistas, making them inviting to the public.

"Multnomah Falls," the sign up ahead read. There are dozens of waterfalls in the gorge, more of them than anywhere else in the Northwest. Multnomah tops them all, about the height of the highest man-made waterfalls Mac had seen, Shasta Dam, Grand Coulee Dam, Dworshak Dam. Man-made waterfalls: water falls over the spillways of dams sometimes, and it always falls inside their powerhouses.

Cool air washed Mac's face. The hills fell away, and he looked at the sun, bisected by a narrow cloud, which took out a quarter of it. He watched the sun set for the last time.

He looked to his right, at the Columbia River, running so close that he felt he could reach out and touch it. The course of that river was the course his life had taken. He had made that choice. Soon he would lose sight of the river, where the interstate veered south toward the center of Portland. From there on, Mac was on its own. The family business that brought him to Portland was the death of a sister, whose estate he was settling. His sister was younger than he, and she was not young. Downstream, the course of his journey that day, seemed fitting to Mac.

60. Multnomah Falls. Fed by springs, this year-round waterfall in the Columbia Gorge, divided into an upper and a lower tier, is 620 feet high, the highest waterfall in Oregon. Aurelio A. Heckert, Wikimedia Foundation.

Mac's last odometer reading before arriving in Portland was 8802.0, at Lewis and Clark State Park. When Lewis and Clark saw the migrating salmon fill the Columbia River from bank to bank, they believed that this country would never run out of fish. Today the explorers would be hard pressed to recognize the river they had ridden, and they would wonder what had happened to the endless fish. The rapids that blocked their way are buried beneath the slack water created by the dams, and to Mac the migrating fish seemed well on their way to disappearing too.

PART 3: PUTTING UP AND TAKING DOWN DAMS

PUTTING UP

Let us see how federal hydroelectric dams came into being fifty years ago. These dams, we note, are not typical dams. Most dams in America are privately built and owned; only around five percent are federal. Private dams are built for specific uses; federal dams are regularly built for multiple uses. Recreation is the most common benefit of the nation's dams, accounting for thirty-eight percent of the total benefit; hydroelectricity accounts for only three percent. For our purposes, we assume that the federal dams were multipurpose dams with a major hydroelectric component.

We further assume that the dams were made wholly or partly of concrete, another reason they are not typical. Most dams are made of rock, earth, or both, so-called "embankment" dams; a Corps of Engineers survey in 1975 found that seventy-two percent of dams in America were of the embankment type, only twenty-eight percent of the concrete. Earthfill dams were already built at twice the rate of concrete dams in the 1930s, when the first dams were built on the Columbia River. Profiting from the techniques of earth-moving projects of World War II and the massive interstate-highway project after the war, earthfill dams were built at ten times the rate of gravity dams in the 1950s through the 1970s. Embankment dams, in addition to being economical, are popular because with some exceptions they can be built on earth instead of rock foundations and out of local materials. It should be noted that there is a hybrid, a concrete-faced rockfill dam, which is counted as rockfill, and that there is another hybrid that uses earth-embankment methods together with cement-enriched soil, known as roller-compacted concrete. Because with the exception of Lucky Peak Dam none of the Columbia Basin dams discussed in this book is the earthfill or rockfill type, we pass over them. For hydroelectric dams, concrete is preferred, since gates can be built into it to control the release of water.

By the start of the twentieth century, the mathematical theory of stresses in concrete dams was well advanced. This was important for the design of dams; theory made their design more scientific, and it had profound affect on their appearance. Theory was developed principally for "gravity" dams, the simplest, which depend mainly on their weight to resist the water load. Because the horizontal pressure of the water against a dam increases with depth, a gravity dam has to be much heavier and thicker at the base than at the top to resist the load; a gravity dam necessarily is extravagant in its use of concrete. The theory was also applied to "arch" dams, which are concave upstream, and which depend on the compressive strength of the arch as well as on weight to resist the water load. This kind of dam can be made much thinner than gravity dams, making use of the weight of the water to press the sides of the dam against the abutments; these dams are best suited for deep, narrow canyons.

There are variants of these two basic types. There is the hybrid so-called "gravity-arch" or "thick arch" or "curved gravity" dam, which resists the water load by gravity alone but which derives additional strength from its arch shape; these overbuilt dams are appropriate for large flows and again for deep, narrow canyons, and they have an aesthetic appeal that straight dams do not. There is the "multiple-arch" dam, which instead of a single arch from bank to bank is made up of a series of arches; like the single-arch dam, the multiple-arch dam requires much less material than gravity dams. Finally, there is the "buttress" dam, which is supported by structures at the toe of the dam on the downstream side, and which usually has a sloping instead of a vertical upstream face; it uses its gravity to resist the load, but it can be made thinner than a gravity dam because of the buttresses. A multiple-arch dam is a kind of buttress dam since its arches are secured to buttresses extending to the foundation. The concrete in arch or buttress dams is commonly reinforced with steel, whereas in gravity dams steel reinforcement is used, for example, in piers (see Illustration 73) or in the base of the dam to secure it to bedrock. American designers were familiar with all of these types of dams.

Arch and multiple-arch dams have the practical advantage over gravity dams of cost, and it would seem that the federal government which is normally interested in saving pennies ought to have preferred them. The Bureau early in its history built several large arch dams and one large multiple-arch dam, and some significant dams of these types were built by private enterprise, but the Bureau and the Corps of Engineers both generally favored gravity dams, the Bureau often the gravity-arch variant. Gravity dams were thought to be better fitted for multiple uses and because of their massiveness to command greater confidence; danger of flooding was always a consideration in their selection. (It made no difference that the one major dam failure in the West, St. Francis Dam near Los Angeles in 1928, which cost over 400 lives, was a gravity dam, not an arch dam.) In making its decisions about dams, economy of materials was not the government's primary consideration, and during the Great Depression one reason gravity dams were in favor was because they employed more workers, though multiple-arch dams too would have required large numbers. The type of dam that the Corps built on the Columbia and Snake Rivers from the 1930s to the 1970s was the same as its first multipurpose dam, Wilson Dam on the Tennessee River in the 1920s: it was a straight, solid-concrete gravity dam with a vertical upstream face and a triangular cross-section, your basic dam. In this section, the dams we look at are of this type.

By the time Mac became a dam engineer, the planning and design of dams and the materials and construction techniques for dams had become largely standardized. That was evident in their appearance, which was more uniform than in the past. The theory of design and the methods of analysis of dams were taught in engineering departments all over. Formulas and rules were laid out in textbooks and manuals. Specialization and organization brought efficiency to every stage of the work, both in the office and at the site. The payoff of standardization was better-built dams. The frequency of dam failures fell off sharply after the 1920s. This said, it was Mac's experience that every dam presented challenging problems that had not been solved before.

To put a dam on the Columbia River, at the beginning there had to be some sort of agitation for a dam at a certain place, and the initiative was usually local. Municipalities, farmers, shippers, and other special interests approached their congressmen in Washington with a proposal. If it was taken up, the dam was entered in a bill, and if it was passed, Congress then approved a fund to get the dam started. In the planning of a dam, the Corps of Engineers or the Bureau of Reclamation first selected a site and made a preliminary design. All kinds of specialists were brought in at this early stage. Materials engineers studied the geology of the site. Surveyors went in with their instruments. Water control engineers and hydrologists determined the flow there. Structural engineers calculated the size of a dam needed to manage the flow. Design engineers came up with a preliminary drawing and artistic sketch of the completed dam in its setting. Economists worked out the costs and benefits of the project. There were still other specialists. Before dams came to be built to generate substantial power, their construction was often denied on economic grounds, but with the dams of the Columbia Basin the return from hydroelectric power almost guaranteed that they would be built.

With Congress's approval of construction and with additional funding, the stage of final design began. The work was again divided into specialties, Mac's hydraulic design one among many. The final design was tested using a scale model of the dam and river, giving designers a chance to see how the dam worked under every hypothetical action of the river, and giving contractors a chance to see what they were bidding on. For dams in the Columbia Basin, final design and model testing usually took about two years.

The construction of a dam – the realization in materials of drawings from the drafting tables – required still more funding. The main raw materials were concrete and steel reinforcement. Work proceeded in stages, which had to be coordinated with one another and with the seasonal river flow. This was the trickiest part. All the time the work was in progress, the river had to be kept flowing, the fish had to be kept moving, and the boats had to be passed.

To get started, earth abutments might be built out a ways from the shore, but at some point the riverbed had to be made dry for excavating, mortaring, and grouting the bedrock to prepare it for pouring concrete. The technique used on the Columbia River dams was "cofferdams," which the Corps had introduced at the beginning of the twentieth century to raise the battleship "Maine." This was a money-saving alternative to excavating a new, temporary

61. Cofferdam Cells and Temporary Fish Passage. John Day Dam. US Army Corps of Engineers.

62. Excavation. John Day Dam. US Army Corps of Engineers.

channel for the river to flow through while the dam was under construction.

Cofferdams are temporary dams enclosing dry areas of riverbed. The type used for Columbia River dams is a series of linked cells, each of which is a hollow tower at first, usually a cylinder, maybe sixty feet across and as many feet high, and normally made of interlocking steel pilings driven down to bedrock. The cylinders are then pumped dry, and filled with sand, gravel, and river dredging for stability. Once a section of the riverbed is dried out in this way, construction of that part of the project begins. While this section is worked on, the river is diverted through other sections. Much design goes into the ultimately dismantled cofferdams and related structures such as temporary navigation channels. All of this work is invisible in the completed, working dam. What is visible are the permanent structures constructed within the cofferdams: dam, powerhouse, lock, and fish ladder.

MAC'S WORK

Other than for a binder of photographs of dams, Mac kept a few, a very few, relics from his life as an engineer. These were a number of loose group photographs of engineers, the telegram from the Bonneville Project that started him on the path to becoming an engineer, a ticket to the John Day Dam dedication, a certificate of service, a cartoon sketch of him on a motorcycle, several snapshots of an office party from the time he retired, a letter from the supervisor who taught a class on the rudiments of hydraulic design, and, of relevance to this section on Mac's work, two reports, one based on a term paper he prepared for that class, and both dealing with the design of spillways. We begin by looking at Mac's work on spillways.

Spillways are essentially safety valves that allow for the passage of excess flow when the river is high. Mac's term paper is about the design of one kind of spillway, known as "side-channel." This structure, a weir, is placed just upstream from the dam, and oriented at right angles to the dam, over which excess water flows into a side channel that conveys it to the stream below the dam. It is useful in narrow canyons where space is cramped and where foundations or costs rule against the common spillways inserted into the dams. The eight-year-old Bureau of Reclamation Hoover Dam – it was still called Boulder Dam when Mac wrote his paper – had a side-channel spillway, and because of the unprecedented problems connected with it, this spillway was of ongoing interest to hydraulic designers. None of the dams on the Columbia and Snake Rivers would use side-channel spillways, and I am not aware that Mac had occasion to design any for dams elsewhere, but his report was used in the Corps of Engineers' revised manual on spillway design twenty years later, and it was placed in the research library of the Corps' central Waterways Experiment Station in Vicksburg, Mississippi.

Mac attached to his report a reprint of an article from an engineering journal giving an analysis of the Hoover spillway done by a Swiss hydraulic laboratory. This is the place to say something about laboratories. They were already in common use in Europe when the Bureau of Reclamation, anticipating unfamiliar problems in the design and construction of Hoover Dam, started its laboratory in the early 1930s. The Corps of Engineers started a hydraulic laboratory for Bonneville Dam at about the same time and for the same reason, to help with problems in the design and construction of Bonneville Dam.

The second report, issued in 1942, arose from extensive laboratory investigations carried out at the Boulder Canyon Project to aid in the design of "overflow" spillways of "overfall" dams. Mac's marginal notes and attached draw-

63. Bonneville Hydraulic Laboratory, Mid 1940s. Engineers from the Portland District and Division offices met with a board of engineering consultants to witness model tests of side-channel spillways for dams. Mac is in the second row, far left, his preferred location in group photographs. US Army Corps of Engineers.

ings were dated by him 1946, around the time he was brought into the little "Hydraulics and Power" group in the Portland District. The main object of the report was to determine the shape of the crest of overflow spillways. The data were obtained from tests on flows over a sharp-crested weir two feet in length, a model of the upstream face of an overflow spillway. The spillways at Boulder Dam are located twenty-seven feet below the top of the dam, one on each side of the dam, and when water in the reservoir rises to that height, it drops over the spillways and is conducted through sloping tunnels into the Colorado River below the dam. Spillways on dams that Mac would design for the Corps in the Columbia Basin differ from Boulder Dam's, but they nevertheless can be modeled similarly, with the help of experiments with sharp-crested weirs. Mac's freehand sketches of the profiles of the kind of dams he dealt with are reproduced in this section as Illustrations 64 and 68, which make the point.

Mac used the Bureau of Reclamation data to plot the shapes of the undersurfaces of flows over the weir for different velocities of approach and for vertical and sloping orientations of the weir, and also to plot the coefficients of discharge for different heights of flow. The report he worked from was prepared by a Bureau of Reclamation engineer. We learn from it and from the report based on his term paper that the Bureau of Reclamation's Hoover Dam and the Bureau's laboratory experiments played a part in Mac's early self-education as an engineer for the Corps of Engineers. Recall also that his mentor in hydraulic design at the Corps came from the Bureau of Reclamation. We might think of the Bureau of Reclamation as Mac's school of design.

We look at some problems that come up in the design of a spillway. A spillway passes water as needed to keep the river from over-topping the dam and possibly damaging or even destroying it. It has to be designed specifically for each dam. Poor spillway design was responsible for a good many early dam failures including the toppling of South Fork Dam in Pennsylvania in 1889, causing the Johnston Flood, which took over 2,000 lives. A spillway is divided into identical bays by piers – Columbia River dams have around twenty bays – each capped by a gate. When a gate is opened, water comes over the crest of the spillway from the pool behind it and passes along the downstream face of the dam. Excess water, that is, flows not over the top of the dam but over the spillway, which is located a good distance below the top.

A spillway has to meet certain requirements. It has to be wide enough to pass the necessary flow for a design flood. The pressures and heads on the crest of the spillway have to be at acceptable levels. The flows and velocities throughout the spillway system have to be acceptable as well. The design of the spillway has to take into account the slope of the upstream face if it is not vertical, the fluctuating volume and velocity of flow of the river, the height of the dam, the geology of the site, the material of the riverbed, and the soils and slopes of the banks. Economic, environmental, and aesthetic considerations can affect the design as well. A spillway may look like not much more than a sloping wall, but there is a good deal more to it than meets the eye.

The shape of the crest is an important element of spillway design. On the Columbia River, the upstream upper face of every dam is a vertical weir, a simple shape, conducive to stable flow, as found by laboratory tests. For stability of the dam the downstream face is slanted, wider at the base than at the top, and to minimize damage from erosion and cavitation the cross section has the shape of an elongated S, or "ogee," curving away at the base toward the horizontal. The curve at the top of the S, the profile of the crest, is designed to follow the under-surface of the sheet of water that flows over the weir, or "nappe." A "nappe-shaped" crest is efficient and protective of the concrete, and for these reasons it is the preferred crest.

We see from Mac's sketch that the nappe forms a jet that first rises above the weir, then falls forward at an angle along the face of the spillway, and finally

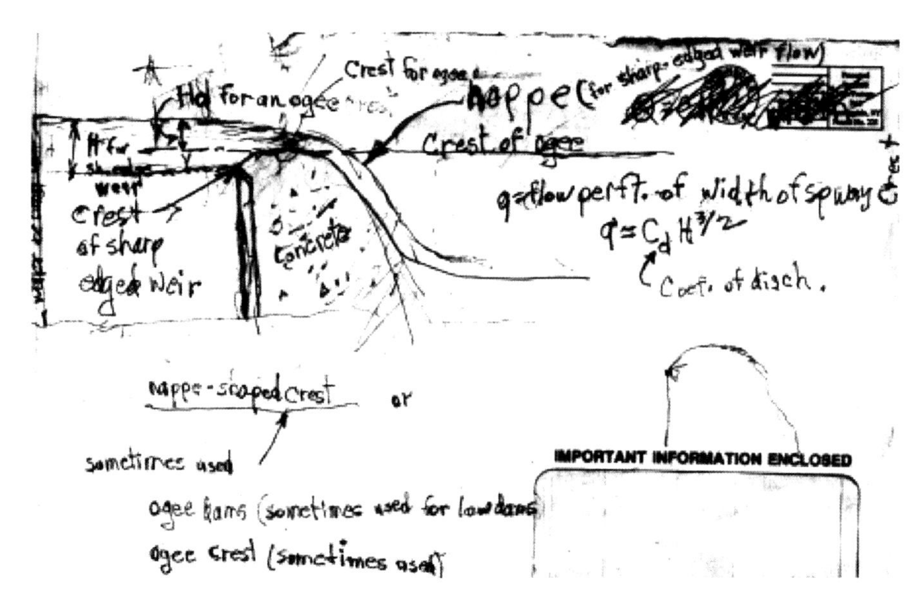

64. Nappe and Spillway, Mac's Sketch. Profile of a Columbia River spillway, the top of the dam not shown. The upstream face is a vertical, sharp-edged weir, which is a type of overflow spillway, as discussed on pages 113 and 114. Mac drew this on an envelope when he discussed his work with me. It shows how the nappe and the spillway fit, ideally hand in glove. It also identifies technical terms used in the text: crest, sharp-edge weir, and ogee spillway profile. The symbols are standard: H is the head of water over the weir, and H_O is the head above the high point of the lower nappe. Albert L. McCormmach.

is guided by a bucket-like circular surface to make a smooth transition from the base of the spillway to the river downstream. In the design of the crest, the initial rise of the nappe is approximated by a circular or elliptical arc. The longer falling segment of the crest is approximated by an exponential equation with empirical constants; the curve is roughly parabolic. At its Waterways Experiment Station in Vicksburg, the Corps of Engineers has developed several standard spillway shapes.

The requirements on the spillway are best met if its design follows the experiments at the hydraulic laboratory; that is, if the profile of the spillway approximates as closely as possible the path of a nappe flowing over a sharp-crested weir. Here is the place to bring up a behavior of water that can be a scourge of hydraulic structures: "cavitation," the formation of cavities. Vapor cavities form when the pressure of a nappe drops below the vapor pressure of water, as happens when it is moving with sufficiently high velocity, commonly between thirty-five and forty-five feet per second. When cavities are swept along in the moving water they are carried to places of higher pressure, and if it is too high to support the cavities, they collapse, producing very high pressures and powerful acoustic shock waves. If the collapses occurred near the surface of a spillway, they make pits in the concrete, and over time the pits accumulate and

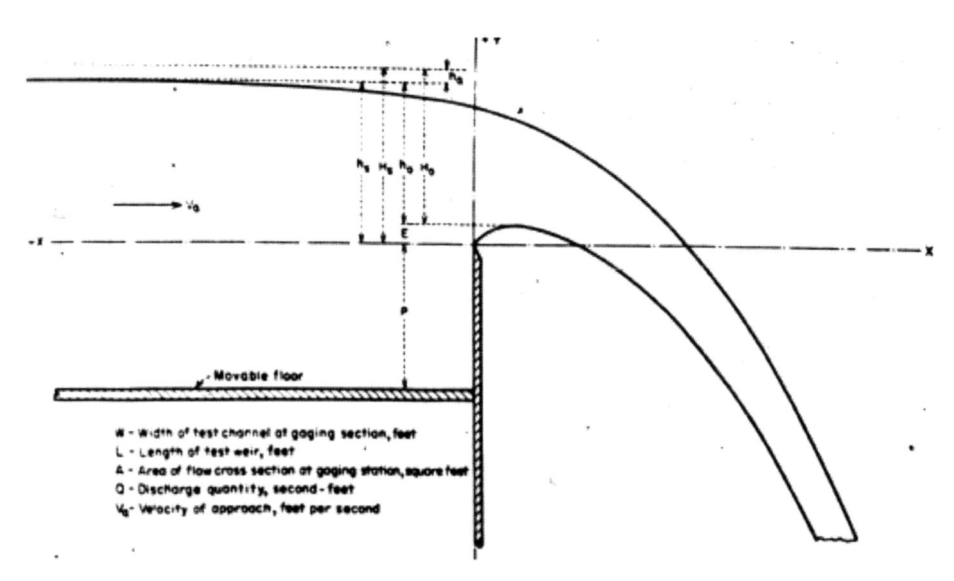

65. Laboratory Model of Spillway Weir, Drawing. The drawing shows a vertical sharp-edge weir, a pool behind it, and a sheet of water flowing over the top called the "nappe." At the Boulder Canyon Hydraulic Laboratory in the early 1940s, experiments were carried out with this model to determine the shape of the under-surface of the nappe for different flows. US Bureau of Reclamation.

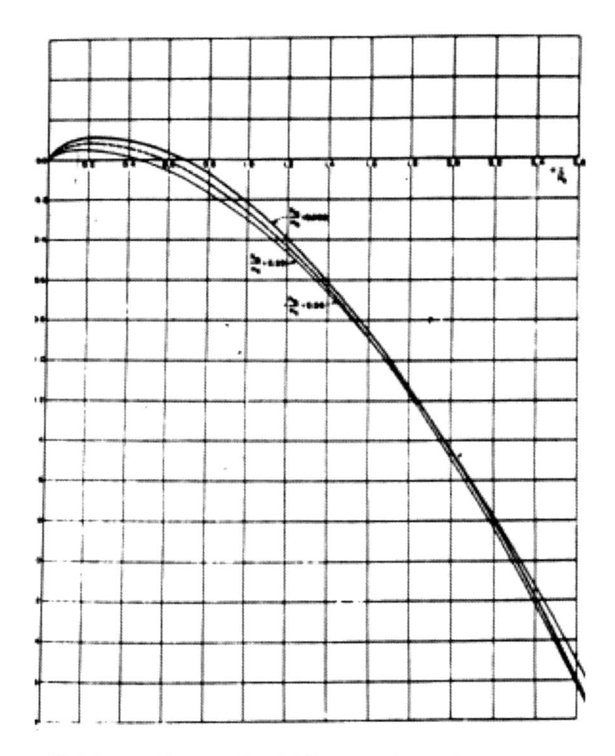

66. Nappe Shapes for Different Flows, Graph. The faster the stream passing over the weir, the flatter is the curve of the under-surface of the nappe. US Bureau of Reclamation.

coalesce, creating bigger holes. Flows moving with velocities beginning around ninety feet per second can cause significant damage to a spillway. It was found that if the flow is aerated, the damage is reduced or eliminated. (The surface of the spillway is modified to allow air to be sucked into the flow, making the water compressible and reducing the pressures caused by bubble collapse.) After having made extensive cavitation repairs in tunnel spillways, the Bureau of Reclamation began in 1967 to modify its spillways to aerate the underside of the flow, and since then aerators have been installed on spillways worldwide. The method of finishing the surface of the concrete also can reduce damage. And so can, of course, hydraulic design, by taking cavitation into account. According to the theory of spillway design, as Mac learned it, if the concrete is fitted exactly to where the water would go if there were no concrete – if it is fitted to the underside of the nappe – the pressure on the concrete is exactly atmospheric pressure, and there is no damage from cavitation, or from erosion (arising from high velocity water flowing over minute cracks in the concrete). That is, if the spillway crest is designed by the book, the water should glide right over the profile of the downstream face of the dam. This ideal can only be approximated in practice, since nappes have slightly different shapes for different ve-

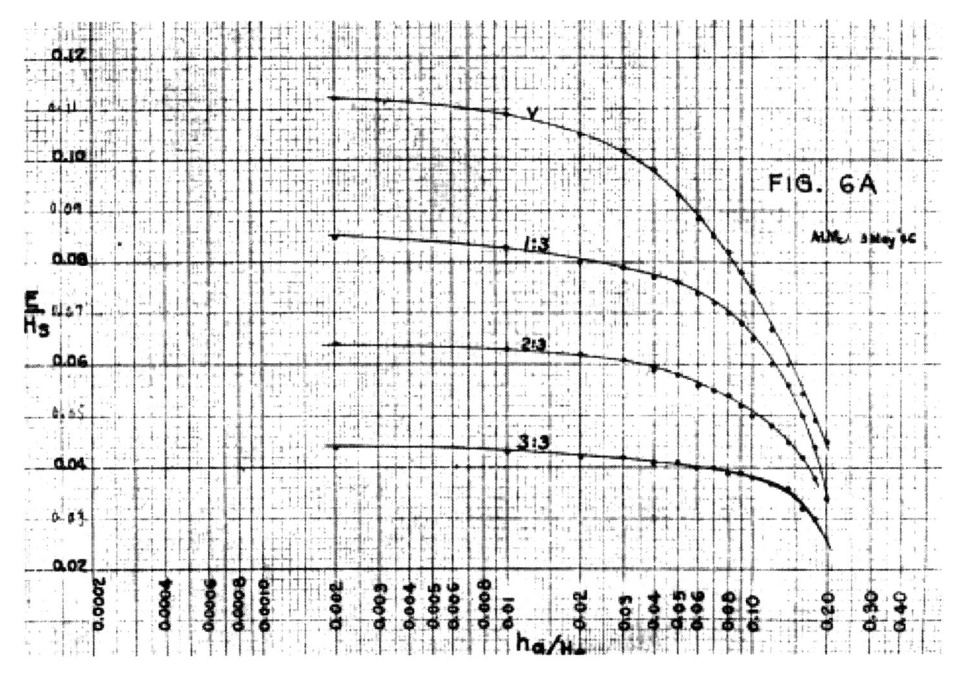

67. Height of Nappe for Different Flows, Mac's Graph. Each curve gives the maximum height of the under-surface of the nappe for different velocities of flow; again, the faster the flow, the smaller is the rise of the nappe. The information in this graph is used to determine the coordinates of the under-surface of the nappe, the profile of the crest. The several curves are for several inclinations of the weir, the top one being for a vertical weir. Albert L. McCormmach.

locities of approach of the water. In the case of a high dam with a negligible velocity of approach, when the water goes over the weir, it springs up higher, because there is no pressure to push it faster. In the case of a low dam with a considerable velocity of approach, the water does not spring up so high. All of this was exhaustively studied using models in hydraulic laboratories. McNary Dam was planned as a low dam with a high velocity of approach, and Mac designed the crest of the spillway with this understanding. Later they learned that the shape of the crest for a high dam works satisfactorily for low dams as well, and a high-dam profile simplifies the design. Tests have shown that the assumed head of water – the velocity of water has an equivalent head – can be exceeded by twenty-five percent without causing harmful cavitation.

The "stilling basin" is another design problem. The water that flows over a spillway acquires a high velocity and energy, which can seriously scour and erode the riverbed and damage the dam. To dissipate a portion of this energy, at the foot of the dam there is a concrete "apron," an extension of the bottom of the S of the downstream face of the spillway. Where a fast moving stream is discharged into a slow moving stream, as occurs below a discharging spillway,

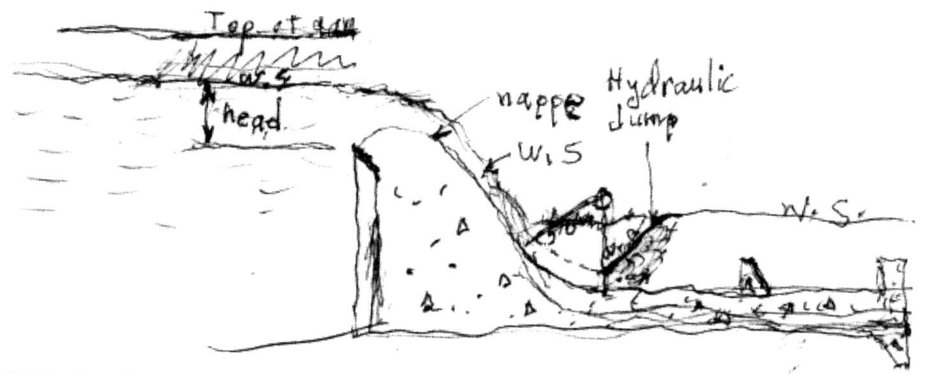

68. Hydraulic Jump in Stilling Basin, Mac's Sketch. It shows the nappe coming over the sharp edge of the upstream face of the spillway, conforming to the shape of the downstream face as it falls, and then rising by a hydraulic jump to the level of the downstream river. A row of baffles on the apron and the end sill are shown. The angle drawn at the foot of the spillway has to do with the curvature of the base of the spillway. Albert L. McCormmach.

the water there rises abruptly in what is called a "hydraulic jump." The flow is bumped up to a higher elevation, say, to the normal river elevation there, and its velocity is correspondingly reduced, and with it its destructive energy. The apron can be made long enough to accommodate the entire hydraulic jump, so that the water exiting it is slowed to the velocity of the river, the desired velocity. It is cheaper, however, to shorten the length by placing obstacles on the apron in the form of "baffles," which by raising the water dissipate the excess energy. Standard baffles are concrete blocks with a vertical upstream face and a sloping downstream face, lined up parallel to the face of a dam. Sometimes there is a second row of baffles, the first causing a hydraulic jump, the second reducing the scouring downstream of the stilling basin. The "end sill," a vertical or sloping rise or step from the stilling basin to the elevation of the river channel, also shortens the length of the hydraulic jump, and by deflecting the high-energy flow away from the channel bed it reduces scouring.

As in the design of spillways, a number of considerations enter the design of stilling basins, among which are characteristics of the flow, slope of the apron, shape of the end sill, presence and shapes of the baffles, side walls of the stilling basin, and exit channel of the river. A lot of care goes into the design, for the

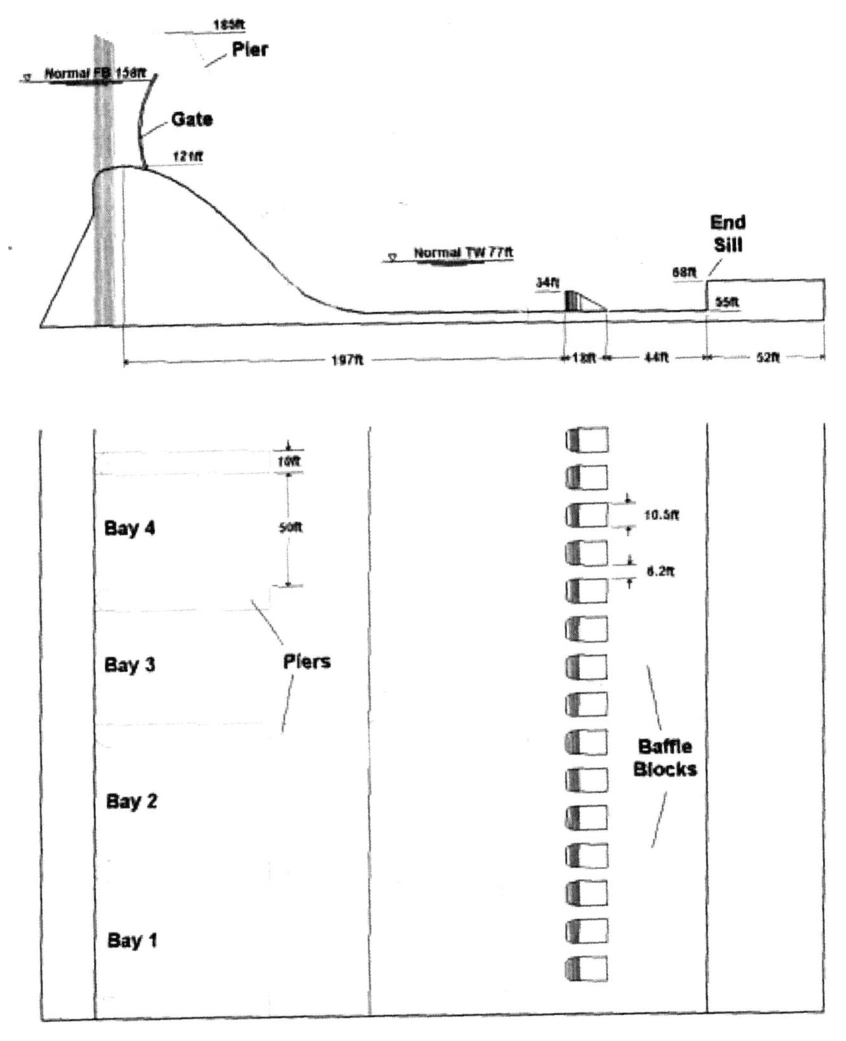

69. Spillway and Stilling Basin, Plan. The drawing at the top is a cross-section of the dam and stilling basin. The drawing below it is a view of the same looking down. The dam is The Dalles Dam on the Columbia River. US Army Corps of Engineers.

safety of the dam depends on getting it right. The plan of the spillway and stilling basin at The Dalles Dam gives you an idea of the dimensions and placement of the underwater structures described here. Baffles there are nine feet tall, the sill is thirteen feet tall, and the horizontal distance from the crest of the spillway to the end of the stilling basin is 311 feet, about the length of a football field. "Flip buckets" are sometimes used as an economical alternative to a stilling basin; these are shaped concrete blocks designed to direct a high-velocity flow away from a dam or other hydraulic structures into a plunge pool, where much of its energy is dissipated on impact.

We turn to the fish ladders Mac worked on. These intricate hydraulic structures simulate the natural rapids of the river at the dams, enabling migrating salmon to surmount the series of 100-foot barriers in their progress upstream. Pools formed by regularly spaced weirs in the ladders provide them with a temporary resting place on their passage over a dam. The pools are good sized; the first ones built at Bonneville were sixteen feet long, forty feet wide, and six feet deep. Known for their strength and bursts of speed, salmon can leap six feet above the surface and even further over the surface, and at first they were expected to leap over the weirs in passing from one pool to the pool above. But leaping tires them and they risk injury, so it was decided to make openings in the weirs a few feet below the surface and to install water jets to get the fish to swim through them.

70. Fish Ladder. At John Day Dam on the Columbia River, Washington shore. US Army Corps of Engineers.

71. Fish Ladder Hydraulics. The author.

Because migrating fish are guided by moving water, the ladders have to pass a lot of water to ensure that the fish find them, and the velocity has to be sufficient to keep the fish moving but not so great as to exhaust them or wash them back downstream. No one knew how much water was needed to get the fish to move. The Fish and Wild life Service thought they knew, but they did not really. They reasoned that the greater the flow, the better, and they asked for as much as they thought the Corps would go for. When the level of the water at the entrance to a ladder rises, as it does from time to time, the lower part of the ladder becomes submerged, but the same velocity has to be kept in there. From upstream, a certain amount of water flows into a ladder, but it also needs to be introduced all along the way with pipes and control valves. The specified drop between each step of the ladder, or weir, is one foot, and the Corps has to agree to it. In the beginning, the water introduced in the lower part of the ladders was taken from the reservoir behind the dam, the headwater. But when you take it from there, you lose water that could generate electricity, so it is not free. Later it was realized that you only have to pump water from the bottom of the dam two or three feet to maintain the specified flow at the entrance, and it is much cheaper to do it this way than to take it from the top and lose 100 feet of head. In addition there is not the excess energy of water falling from a height, that otherwise has to be dissipated at the bottom to prevent cavitation damage. The ladders have competition from the water that flows out of the draft tube outlets of the powerhouse, which fish are naturally attracted to. To discourage them from trying to go in there, a collection channel is installed across the front of the powerhouse and supplied with auxiliary water to divert the fish to the ladders. All of the dams the Corps of Engineers built on Columbia and Snake Rivers were fitted with fish ladders, at the beginning one at each shore, though later it was found that a ladder on only one shore is sufficient. The Corps' design for fish ladders was new, and it became the standard the world over.

When Mac went to work as an engineer, hydraulic designs were drawn with drafting instruments on drafting tables, calculations were made with slide rules, logarithm tables, and mechanical calculators, and the operation of dams was done manually. Today these things are done with the help of computers. Together with model-testing in hydraulic laboratories begun in the 1920s and 30s, computers from the 1960s have revolutionized the analysis of dams.

The changeover to computers began in the middle of Mac's career, and it came about hesitantly. Mac was one of the first engineers in his district to work with a programmer to solve an important problem with the computer. The problem lay outside his specialty, hydraulic design, though it was relevant to it. I bring it up because it is the one occasion in which I had a part, a very small part, in Mac's work, and for this reason I know rather more about it than I do about some other work of his. It shows how he went about his work, with determined persistence.

In the planning of dams, you have to know the water elevation below the dam for various flows of the river. This has to be calculated. You start with a given water elevation at a section in the river, and then you find the elevation at another section back, say, a quarter of a mile. You find it by assuming various backwater elevations until you hit the right one. You know it is right when the energy balances: the energy of the flow downstream plus the friction loss between the two sections equals the energy of the flow upstream. Then you go another step up the river and do the same thing. The computation makes use of years of data on elevations for Columbia River floods. When you have computed enough of these backwater elevations, you have the river profile for the assumed flow. When Mac started out, backwater elevations had to be figured by hand, using mechanical desk calculators. The planners might have used only two or three backwater river profiles because the calculations took so much time, weeks, even months. Later, when it was done by computer, they might have sixty or eighty profiles. The design of a dam came out so much better because they had all this additional information, and they obtained it in short order. This saved millions of dollars in the design of dams and in their operation.

When I was going to college, I worked one summer at the Walla Walla District Corps of Engineers. The idea of electronic computers became big at just this time. These machines were new then, and no one in the district had taken a problem to them. The supervisor of my section, hydrology, had heard about them, and he was curious to see if they did anything useful. He sent me to Portland where IBM had an office to find out. I described the notorious backwater problem to a computer programmer there, who on the spot drew a flowchart, the scheme of a solution. I gave the chart to my supervisor, who filed it in a drawer, down deep, where it sat.

Not too long after that, but after I had the left, the district picked a dozen of its engineers to go to Portland to take a course on computers that IBM was offering free. Mac was one of those it picked. Every engineer took a problem with him, and Mac chose the same problem as mine, backwater. He knew that I had gone to Portland with the problem and that I had brought back a flowchart. He went to my supervisor and asked him for it. Reluctantly, my supervisor pulled the flowchart from the drawer and gave it to Mac on loan, with Mac's promise to return it. Mac's own supervisor, an old backwater hand, told him that computers no doubt are a wonderful thing, but there is one thing that computers will never do, compute backwater profiles. Actually, they are ideal for that. The backwater problem is all cut and try. The instructor for the Portland computer course was the same IBM man I had taken the backwater problem to. He had a strange teaching method. He kept telling the class how dumb it was, the dumbest class he had ever taught. A few engineers got discouraged and went home, possibly the weeding out the instructor intended. He had everyone pick a simple problem, probably not the one he brought but a simpler one he had come up with during the course. Mac bought a big brown sheet from a butcher and laid it out on his hotel-room floor, and at night while he was taking the course he worked out the flowchart for the simple problem he had picked. When it came time to put the problems on the computer, everyone who remained in the teacher's dumbest class put his problem right through. The business of putting it through was coding it. With Fortran, the machine did the coding, but it had not been invented yet. Everything, instructions and data, had to be written in machine language, in decimal arithmetic. You had to tell the machine exactly what to do, and you had to know tricks to do it. Mac's simple problem went through all right, but the backwater problem was completely beyond him.

Next the district brought in a man from IBM to spend two weeks there to see if he could solve some useful problems. He spent all of his two weeks working on one engineer's problem, the design of a spillway gate. This engineer did not have a flowchart, and so the IBM man had to start from scratch. On the IBM man's last afternoon he came to see Mac. He was amazed. Oh, you have a flowchart! The IBM man sat down and wrote out the code, exactly following Mac's flowchart. That was fine, except that he made a mistake, and Mac had to revise the coding.

Mac took some people from water control to Portland to show them how to use the backwater program. The computer was an IBM 650, then on the market for only two or three years but already becoming the workhorse of business. It occupied a big room, bigger than the floor space of Mac's house. Because it ran with vacuum tubes, there had to be blowers to get rid of the heat, and these were noisy. The computing was done on a magnetic drum, which had a memory of 2,000 ten-digit words. The drum made 12,500 revolutions per minute, causing a big hum, and the consul had a lot of blinking lights. When the machine was running, you knew it; it was a performance.

The computer was much faster than hand calculators, but the backwater problem still took a lot of time. This was because the bed of the river is irregular. To calculate the friction loss of energy, roughness coefficients had to be assumed. You take the roughness coefficient for the assumed elevation upstream and the roughness for the elevation downstream, and you average the two, and that allows you to calculate the average friction loss between the two sections. The way Mac originally programmed the problem, he would assume a value for the coefficient and then run the profile, then assume another coefficient, and run another profile. But every time the coefficient changed, it took up time, and he was running this program on an expensive computer. He was spending more time changing coefficients between every run than the machine took in the computation. (Early computers worked with punch cards, remember.) So, obviously, that was not the solution. Then the idea came to him. The computer had a peculiar feature called table lookup. You would give the machine instructions for table lookup, and provided you had put a table in there, it would go back and read it. So Mac plotted curves for the coefficients. He had the data because over the years, the Corps had determined values of the coefficient at different elevations of the river. Every time the elevation changes, the coefficient changes, and so there is a coefficient curve for different elevations for each section. Mac redesigned the program so that the data he put in would give him a coefficient curve for each section. The coefficients did not have to be changed by hand any longer, it was completely automatic. Mac never actually used the program himself, since backwater was not his problem and he turned it over to other people. For a while, his program was the only one there was. Backwater computations were a man-killing job everywhere there were dams, and Mac got letters from all over the country, even from foreign countries, asking about his program. It saved thousands of man-hours.

We end this section on Mac's work with several photographs giving examples of another class of problems, complications which arise in the operations of dams. Mac frequently made inspection trips to sites to decide what action to take when one of them came up.

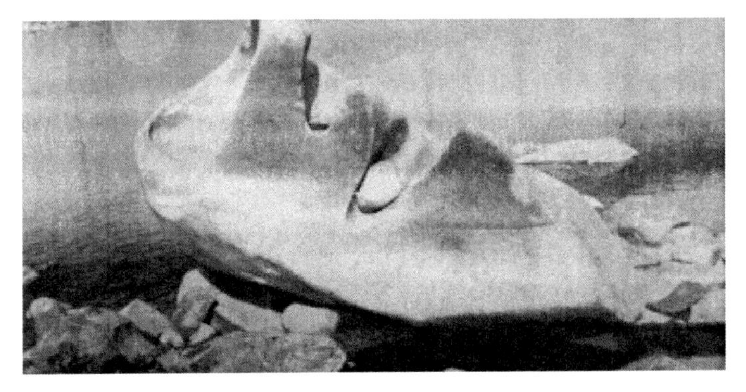

72. Erosion. An example of erosion of hard matter by water at John Day Dam. Mac took the photograph of this Henry Moore-like form, sculpted by natural forces.

73. Erosion. Two kinds of erosion are frequently encountered at dams. One is cavitation, which is discussed in the text; the other is abrasion, which results from the circulation of rocks, gravel, and other debris over a concrete surface. The two sources are readily distinguishable: erosion from cavitation is pitted, and erosion from abrasion is smooth. Abrasion is the type of erosion shown in this and the next photograph. We see the concrete stripped away down to the reinforcing steel on pier 10 at John Day Dam. Piers are vertical structures that separate the bays of a spillway. Albert L. McCormmach.

74. Erosion. Similar damage, closeup, Pier 13. Albert L. McCormmach.

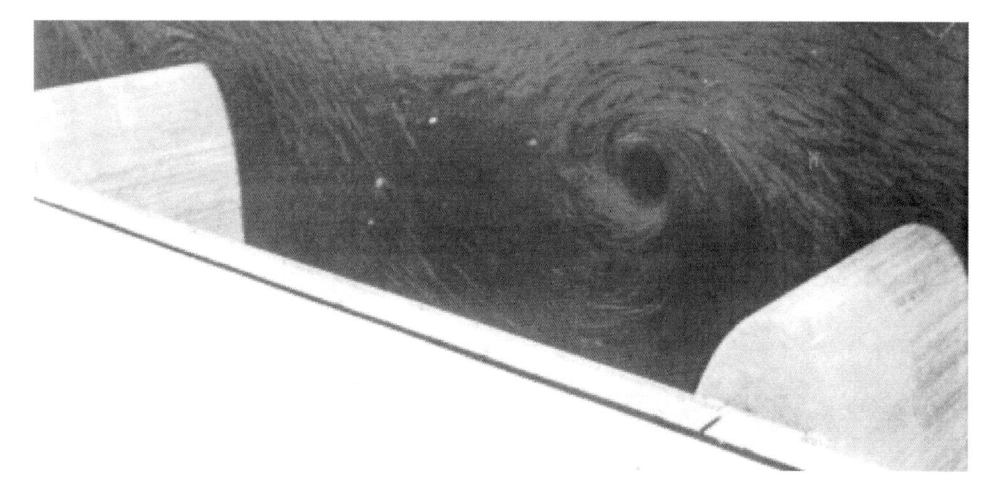

75. Vortex. Vortices like the ones shown in this and the next two photographs form upstream from a bay and pass over the spillway. They can impede the operation of a dam and cause a higher mortality in the juvenile fish coming over the spillway. Vortices contain a lot of energy, and a technology has been invented to develop power from vibrations setup by vortices in a river, currently promoted as an alternative to dams in a river. The vortex shown here is at bay 16, McNary Dam. Albert L. McCormmach.

76. Vortex. Flow through McNary gate 16 with adjacent gates closed. Air from a vortex can be seen on the left. Albert L. McCormmach.

77. Vortex. Another view of the vortex at gate 16, numbered from the left. Albert L. McCormmach.

TAKING DOWN

Dams look as if they are meant forever, but they are not. They take a beating day and night, in every weather and under every river condition. Over time they develop problems, and for a while the problems get fixed, but somewhere down the line the question of replacement or removal can come up. Most American rivers filled up with dams a long time ago, and some of these have outlived their purpose or have been abandoned, while others have reached or are approaching the end. Safety, cost effectiveness, and, increasingly, the environment are key considerations in deciding what to do with them.

It is estimated that at least 465 dams were removed within America between 1912 and 1999, an average of five per year, and so their removal is nothing exceptional. However, when this number is compared with the total number of dams today – the Corps of Engineers estimates that America has around 75,000 dams over six feet high and countless thousands smaller than these – they are still rather few, leaving the fate of most aging dams still to be addressed.

The frequency of dam removals is increasing. Most have occurred in the last quarter century, with nearly twice as many in the 1990s (177) than in the 1980s (92). The increase can be explained partly by the increase in the number of dams that are fifty or more years old, partly by the wear and tear accompanying aging, partly by better record keeping, and partly by the environmental movement, which after having effectively stopped the construction of dams made their removal the next goal. A coalition of river preservation groups took advantage of the many upcoming hydroelectric-dam license renewals in the 1990s to press the government to adopt a decommissioning policy. The government proved responsive to this latest environmental cause celebre; Bruce Babbitt told a Trout Unlimited symposium in 1994 that he "would love to be the first secretary of the interior in history to tear down a really large dam." (He took out small Edwards Dam in Maine in 1999, the first federally mandated environmental dam removal in the country.)

That same year, the agency of Babbitt's department responsible for approving the construction of hydroelectric dams took on the additional responsibility of deciding about the removal of hydroelectric dams. The Federal Energy Regulatory Commission (FERC), formerly the Federal Power Commission, issues hydroelectric licenses for up to fifty years, under conditions complying with the Endangered Species Act and other environmental laws. In response to a Supreme Court ruling, in granting and renewing licenses the commission gives "equal consideration" to environmental issues and power production. The commission looks to the National Marine Fisheries Service and other agencies to determine the environmental conditions that a dam must meet before its owner's application can be approved. In the case of an existing dam, when the owner's hydroelectric license expires the commission can now order the dam to be decommissioned at the owner's expense if the environmental conditions are not met.

What it means to "decommission" a dam needs clarifying. Where a dam is partially or completely removed to help restore the natural condition of a river, it is said to be decommissioned. Partial removal is sometimes preferable to complete removal. In the case of partial removal, the flow of the river is restored, while what remains of the dam holds back any sediment accumulated behind it, which if released could cause damage downstream and affect the quality of the water. "Notching," making an opening at the top of a dam for water to pass through, is an example of partial removal. Other examples include creating an opening at the bottom of a dam and removing the abutments of a dam. More costly than partial removal, complete removal restores the river to its full natural flow. Alternative expressions for decommissioning are "breaching," "taking down," and "removing." I use "removing" to stand for both partial and complete removal of a dam.

The methods for removing dams are straightforward: pneumatic drilling and blasting in the case of concrete dams, and excavation and hauling in the case of earthfill and rockfill dams. During removal, the reservoir water and stream are temporarily diverted; cofferdams are sometimes used for this purpose. A demolition site can resemble a construction site. Whatever method is used to remove a dam, it entails a major construction project, and a major expense.

Like the construction, the removal of dams carries economic and social risks, which cannot be entirely foreseen. Inevitably a number of interested parties have to be consulted, usually calling for compromises and often for compensations. The release of impounded water has a range of physical and environmental consequences, the most serious of which is the flushing of sediment downstream. The sediment that is not flushed away can cause problems too, since it may contain toxic wastes. There are other kinds of damage. The now-exposed basin of the drawn-down reservoir can look unsightly; it might, for example, have a forest of tree stumps. Riverfront property alongside the reservoir no longer fronts the river, with a loss of appeal and value. Types of recreation are sacrificed. There are problems with water habitat and wetlands. Fish are put at risk, even if their welfare is the reason for the removal of the dam in the first place; since their natural spawning grounds have been destroyed by the dams, their removal can do more harm than good in the short run. Removal of dams is a disturbance of nature comparable to that of putting them up, and it has to be carried out with comparable foresight and caution. Although the methods of taking down dams are widely known, we are only just learning how to use them properly. At this stage we know better how to put up dams than to take them down.

Because few big dams have been removed, there is not a large body of experience to draw on, and there can be major disagreements. A case in point is the projected removal of PacifiCorps' Condit Hydroelectric Project on the White River, a tributary of the Columbia River in Washington. Its license has expired, and the environmental conditions of its renewal include the addition of fish ladders, which would cost the owner more in construction and lost revenue than its alternative, the removal of the dam. The company has applied to the FERC for removal, and the National Forest Service and most environmental groups favor this course, seeing it as a precedent for the removal of other big dams. What stands between the FERC and PacifiCorps is their different proposals for dealing with the major problem of accumulated sediment behind Condit Dam. The FERC wants to dredge the sediment or to build a bypass around it. PacifiCorps wants to flush as much of the sediment as possible down the river quickly to shorten the damage to marine life. If the sides come to agreement soon, (before Giles Canyon Dam is removed, see below), the 168-foot-high Condit Dam will be the biggest dam so far to be removed.

78. Condit Dam. Candidate for dam removal. GEDL.

The removal of a dam does violence to a river and to the fish and other life, but a natural river is a resiliant system, and in the long run it is better off without the dam. This is the essence of the environmental case for dam removal. PacifiCorps' plan for removing Condit Dam shows how it works. A big tunnel would be cut, drilled, and blasted through the concrete base of the dam. The water backed up by the dam would drain through this opening in about six hours, carrying with it a large quantity of sediment. A new, narrower channel would be formed upstream. Sediment that had not been flushed would gradually erode until eventually vegetation would take root and stabilize the new channel. The flushed sediment would have a negative impact on the resident fish for a couple of years, and then there would be an improvement. The remaining dam would be cut into blocks to be recycled there or moved somewhere else for disposal.

A series of big dams in the Pacific Northwest are either scheduled or being considered for removal. All of them are on rivers with salmon runs, and in each case the cost of providing the required environmental protections for license renewal outweighs the economic returns. Removal is in prospect for the lower four of six aging dams on the Klamath River in southern Oregon and northern California. Built between 1908 and 1962, they range in height from 128 feet to 463 feet. Three of the dams do not have fish ladders, and the fish facility at the fourth is inadequate; further, the dams are responsible for toxic algae buildup in the river. Under pressure from conservation groups, tribes, commercial fishers, and Oregon and California governors, in 2008 the owner, again PacifiCorps, signed a nonbinding agreement with the states and the federal government and other stake holders to remove its dams. The Bush administration made an exception to its opposition in principle to the removal of hydroelectric dams in the Columbia Basin, seeing it as the only way for Klamath fishers and farmers to live in peace. The removal would begin in 2020 (if not delayed), far enough in the future for an alternative source of power to be found to replace the power from the dams, currently sufficient to supply 70,000 homes. The delay would also allow for scientific studies to be made of the environmental effects of releasing the sediment behind the dams. The Klamath Basin once had the third largest salmon migration on the West coast, and if the dams are taken down as planned, it will open up over 300 miles of spawning habitat. The project will be the largest dam removal in America.

The decommissioning of another four dams in another basin in the Pacific Northwest is a certainty. Under various ownership and with varied uses, these dams are located on the Rogue River and its tributary Elk Creek in southern Oregon. Elk Creek Dam was authorized as a Corps of Engineers project in 1962, but formal construction began only in 1986, by then late in the game. By court order the Corps discontinued construction the next year. For years after, the Corps trucked the fish around the unfinished dam – it was built to only a third of its planned height – before deciding that it would be cheaper to breach the dam. It was delayed for several years in carrying out its plan because of local opposition to the removal. When at last the Corps set off the first of a series of blasts at Elk Creek Dam extending from July to September, 2008, photographs of flying concrete were carried by the media, giving heart to advocates of restored fish runs. By removing material from the spillway and an abutment, notching not destroying the dam, the Corps allowed for the possibility that one day it would return to complete it. Next to the Columbia, the Rogue had the largest salmon runs in Oregon, and when all four dams are down, for the first time in over a century the salmon will move unimpeded along a 157-mile waterway from the ocean to the Cascades.

79. Elk Creek Dam Blast. The start of dam notching by the Corps of Engineers, 15 July 2008. Waterwatch. Savage Rapids Dam was removed in 2009, Gold Ray Dam in 2010.

Removal is also a certainty for two large concrete dams on the Elwha River in the Olympic Peninsula in Washington, off the Strait of Juan de Fuca. Privately built between 1914 and 1927, for years 105-foot Elwha Dam and 210-foot Glines Canyon Dam generated hydroelectric power, stimulating economic growth in the peninsula, but now the combined power generated at both dams provides only one third of the electricity needed to run one paper mill. When the Glines Canyon Dam license came up for renewal, environmentalists began their long campaign to have it and Elwha Dam removed. By an act of Congress in 1992, the federal government acquired the dams for habitat restoration. The interior department in agreement with the environmentalists decided that the only way this could be done was to remove the dams. This was a big decision for the fish, which had been reduced to pitiful numbers. Salmon which once had access to about eighty miles of habitat in the Elwha Basin and which had returned in numbers of around 400,000 had been reduced to fewer than 4,000 with access to only five miles of habitat. The removal of the dams is scheduled to begin in 2111, the biggest removal up to then.

80. Elwha Dam. Scheduled for removal. If for no other reason this dam should be removed because it is so ugly. Rafal Pocztarski, Wikimedia Commons.

The examples above all concern newsworthy big dams. Small dams are easier to remove physically and logistically, and they are more typical; the average height of dams removed in America in the twentieth century was about twenty-one feet, the median height about fifteen feet. Examples of successful small-dam removal in the Columbia Basin are Lewiston Dam, a thirty-nine-foot, obsolescent, hydroelectric dam on the Clearwater River in Idaho, removed in 1973 to improve the salmon run, and an eight-foot replacement dam built by the Corps of Engineers in the course of some work on levees on the Walla Walla River, removed in 1997 to provide passage for migratory and resident fish.

The removal of big federal Columbia Basin dams has not been tried, but the Corps has studied the benefits for juvenile salmon of breaching the four Lower Snake River dams. The Corps would do this by removing the earth abutments of the dams, leaving standing in the middle of the river the concrete monoliths. People may come to accept this as the lesser of evils, but it is an ugly solution, and costly. The electricity from the four dams is only a fraction of the electricity from the whole system of dams on the Columbia, but it is very considerable all the same, sufficient to power Seattle, the Pacific Northwest's largest city, and nearly equal to the entire output of the TVA dams. Together their output equals the combined output of John Day and Bonneville Dams. If all of the big dams on the Columbia and Snake Rivers were to be taken out – an eventuality almost certain not to happen – the economic impact on the Northwest would be catastrophic.

The desire to see dams removed from American rivers is a composite of desires, which have similarities with the desires to put them up in the first place, only the objects of the desires take on different meanings. Recall the desires. Desire to build: this takes the form of a desire to rebuild; for a half century, since the publication of Rachel Carson's *Silent Spring*, Americans have come to recognize the destructiveness of some of their ways of doing things and the importance of allowing nature a breathing space to rebuild the ecological health of the planet. Desire for earthly paradise: humans living in harmony with nature is one kind of paradise, a community of irrigated, electrified family farms is another, though they are not incompatible. Desire for multiple river uses: reorder uses to favor those that nature intended for the river, including natural processes not directly linked to human uses. Desire for river passage: it is the fish's turn again. Desire for safety: the collapse of a big dam is as great a disaster as a giant flash flood. Desire for autonomy: the analog to the economic development of the basin is the ecological integrity of the watershed. Desire for security: in our time, which is under a new threat, terrorism, dams are a vulnerable target. The other earlier desires – desire for power and desire for economic growth and recovery – are either beside the point or are denied in the case of dam removal. Of course, persons who make decisions about removing any particular dam have to weigh costs and returns, adhere to the law and respond to court rulings, and generally consider all sorts of practical matters.

There are desires specific to dam removal. There is a desire, for example, for a more equitable world. Big dams have not lived up to their promise to improve the conditions under which most people in the world are forced to live. Benefits have fallen disproportionately into the pockets of a minority of the already well-off. Inequality is on the rise most everywhere. The daunting problems of poverty and hunger remain with us as ever, and the cost of the unfairly distributed benefits is damage to the land and the water and the living beings that inhabit them. There is a corollary desire for an alternative world view, one not focused on economic growth and mastery of nature, instead one more in harmony with nature, and the removal of dams follows from it. Desire for destruction in its own right is not on the list. Certainly humans time and again have shown their readiness to wreak destruction on a vast scale, and no doubt among advocates of dam removal there are those who experience a thrill at the sight of a demolition, as indeed do many who are not advocates, and there may be persons with such an animus toward the whole industrial drift of the world that they welcome the destruction of one of its iconic symbols, but there is no discernible pattern, and nihilistic passions are quickly spent.

The difference in the desires to put up dams and the desires to take them down has to do with a difference in perspective, that of the region and the nation in the dam-building era, and that of the planet, the web of nature, and humanity in the environmental era. There is a corresponding difference in carrying out the desires. When the dams were put up, most persons could see something coming their way before long, and the opposition was disorganized and ineffectual. Proponents of dam removal today form powerful coalitions, but the benefits of dam removal are a harder sell. To many, dam removal promises more loss than gain. When the dams were put up, there was the river, and now there are the dams and the river, and there are ways of life and ways of using the river that have come about because of the dams. For this reason, it is harder to take down those same dams than it was to put them up.

THEN WHAT

Society can change its mind about a physical power it once welcomed. It can decide to forgo developing it further or to replace or supplement it with other forms of power. This has happened with hydroelectricity in some parts of the world, as we have seen. To understand the implications of this change, we need to place hydroelectricity among the other sources of energy. Over eighty percent of the world's power comes from fossil fuels, of which the most abundant and the one certain to outlast the competing non-renewable sources is coal. At present oil accounts for more of the world's energy consumption than coal, a third verses a quarter, but oil extraction is near its peak, while coal is the fastest increasing fossil fuel; in energy units, oil reserves are estimated to be only about half of the reserves of coal, about the same as those of natural gas. Nuclear fuel follows coal in its share of the world's energy use, about six percent. On the basis of current inventories, according to one projection the earth will effectively run out of oil in forty years, natural gas in sixty years, nuclear material in seventy years, and coal in 150 years. In addition to their depletion, there are mounting environmental objections to each of them. Measures to conserve energy and to use it more efficiently will extend the lifetimes of non-renewable energy reserves, but they cannot replace them.

Renewable resources will supply an increasing proportion of the world's energy. Most of the world's energy comes from the sun, if indirectly. Fossil fuels store it. Hydroelectric energy depends on it: heat from the sun raises water from the seas, which collects on mountains, from where it runs back down to the seas, poweringhydroelectric turbines on its way. The sun's energy is renewable energy, but only a small fraction of it can be captured. Although it would take only 0.02 percent of the energy the earth receives from the sun to replace fossil and nuclear fuels, for practical reasons even this tiny fraction is well beyond our reach. Of renewable resources now, biomass contributes most to the world's energy consumption, about four percent, followed by hydroelectricity at two to three percent. The contribution of solar, wind, and geothermal resources is far less, but because these do not run out and do not harm the environment they receive considerable encouragement, and their share is growing. There are wild cards, notably nuclear fusion on which publicly supported research continues after decades of disappointment. The world today relies on highly concentrated sources of energy such as coal. When it runs short, what will replace them, how much it will cost, and the technology that will deliver the power are all unclear. All we are certain of is that major change is coming.

In terms of energy, we see that hydroelectricity is a bit player on the world stage, even though in some places such as the Pacific Northwest it plays a prominent part. Yet much of the world's power, whatever the source, is converted into the convenient form of electric power, and as a share of the world's electricity, hydroelectricity takes on greater importance. Hydroelectricity makes up fifteen to twenty percent of the total electricity in the world. When Mac joined the Bonneville Project, there was a clear, as yet relatively untroubled, goal, the production and marketing of hydroelectricity, and society had little reason to question it. Hydroelectricity met a good number of basic human desires in the dam-building era, and over the course of Mac's career much that was of value got accomplished. After the era ended, as we have seen, Mac looked back on that accomplishment with misgiving, for the environmental cost in the long run seemed to him too great. With hindsight, he could wish that the accomplishment had never happened.

With Mac's reflections on the development of the Columbia Basin in mind, we will consider the prospects of hydroelectric development in America and in the world at large. We begin with a brief survery of some of the criticisms. Since most hydroelectricity is generated at dams, the criticisms are largely directed at them and at the reservoirs they form. Dams in the developing countries displace tens and hundreds of thousands of people, who are usually poor, and are even poorer after their relocation. Reservoirs destroy wetlands, forests, agricultural land, and biologically diverse ecosystems in the rivers and on the banks. They change the flow and temperature of rivers, with harmful effects on life. In tropical climates, reservoirs breed malaria mosquitoes and other dangerous organisms. Sedimentation in reservoirs affects alluvial plains and wetlands downstream, and it shortens the life and increases the cost of maintenance of dams. Reservoirs reduce flow in the river downstream from the dams. They invite aquatic weeds, reduce the concentration of dissolved oxygen, and build toxicity in the impounded water. They give off greenhouse gases, and if the reservoirs happen to bury forests the quantity of gases is very considerable, in which case they bring into question an advantage hydroelectricity was thought to have over fossil fuels. Dams inevitably weaken and at some point may have to be replaced or removed, a risky proposition for the river and the life it supports. In addition to the damage they do to the environment, dams result in economic waste. They are expensive, cost overruns are the rule, and when everything is taken into account returns frequently fall short of investments. The list goes on and because it is well publicized it is little wonder that many people have has lost faith in hydroelectricity as an acceptable alternative to fossil fuels.

With so many problems caused by dams, can anything be said in defense of hydroelectricity? Let us see what kind of case can be made. To start with, there is a lot of hydroelectricity around the world, and although it may not be ideal, neither is any realistic alternative. According to Vaclav Smil, over twenty countries depend on hydroelectricity for ninety percent or more of their electricity, and nearly a third of the countries depend on it for more than half; they have a big investment in hydroelectricity, and it works for them. There are many countries that need more power or could distribute power more widely, and hydroelectricity is an option for some of them. One quarter to two fifths of the world's peoples have no access to electricity from any source. We find them concentrated in Africa, Asia, and Latin America, continents with the greatest undeveloped hydroelectric potential. Whereas in North America and Europe, a third to one half of the potential of their rivers has already been tapped, in Africa only three and a half percent has, and in Asia it is eleven percent and in Latin America twenty percent. Political conditions permitting, the development of hydroelectric power in these parts of the world could improve the quality of life of the mass of the people, bringing them a step closer to the standards of the developed parts of the world, and if the development is carried out responsibly with only modest environmental costs.

Hydroelectricity has other virtues. Compared with alternative sources of electricity, it is decidedly more efficient, and hydroelectric power plants cost less to run and they last longer than facilities for generating electricity any other way, making energy cheaper to produce and potentially more affordable. Also the ratio of the energy delivered by hydroelectricity to the energy required to build and maintain its facilities is much higher than it is for other sources of energy, including other renewable sources. Hydroelectricity pays for the benefits society receives from the operation of multipurpose dams; their reservoirs meet residential and industrial water needs, control floods, provide for irrigation, and open channels to navigation. Hydroelectricity insures the reliability of a power system; electric power produced by storage projects, in which water is shifted back and forth between a high and a low reservoir in response to power needs, is steady, in contrast to the intermittent generation of electricity from wind and other renewable sources, and because it can be brought on line without delay it is ideal for meeting peak loads in a transmission grid.

Hydroelectricity has two major advantages over electricity produced the usual way. First, although it contributes to global warming, its contribution is relatively minor; studies have shown that where electricity is produced by dams instead of by fossil-fuel plants, smaller quantities of greenhouse gases are released into the atmosphere. Second, hydroelectricity does not run out. The fuel, rivers, is renewed by natural means and produces no waste, and it is free into the bargain. This was recognized from the start. In 1937 Ross, appointed that year as the first administrator of the BPA, gave a statement to this effect to the press: electric power is preferable to its alternative, steam power, since the latter requires fossil fuel, which when it runs out is gone forever. "But a great River is a coal mine that never thins out. It is an oil well that never runs dry. The Columbia River will flow through the Bonneville and Grand Coulee dams . . . as long as rain falls and water flows downhill to the sea." The finite lifetime of the dams aside, this statement is as true today as it was when it was said. Sources of power with the two advantages of hydroelectricity, minimal pollution and renewable fuel, are in high favor today.

Under the threat of global warming, the Obama Administration has supported hydropower at levels not seen since the 1980s, regarding it as one part of the solution to the energy crisis as well as an economic stimulus. The FERC is swamped with hydropower proposals, and although most of them are for making improvements at existing dams, it is noteworthy that the department of energy has identified nearly 6,000 potential hydroelectric sites, representing forty percent of the nation's existing hydroelectric capacity, and the region with the greatest number of sites is the West, Washington the leading state.

All of the ways people have chosen to satisfy their desire for physical power have had unwanted as well as desirable consequences. If alternative, "clean" sources of power, such as sun and wind, should be developed on a large scale, we can expect the same mix. The World Bank has returned to supporting hydroelectricity, pragmatically, on a case-by-case basis, taking the view that there are good dams, and there are bad dams.

Hydroelectricity is unequally distributed over the face of the earth. After China, Canada, and Brazil, America is the next biggest producer of hydroelectricity, and forty percent of America's total comes from the Pacific Northwest. The region enjoys the greatest per capita consumption of electric power, and unlike the rest of America it derives most of it from dams. The proportion is likely to decrease – most of the big dams there have been built, and society is looking at alternatives – but for the foreseeable future, electricity generated on the rivers of the Columbia Basin will provide the largest share of the Pacific Northwest's electric power; today it provides well over half, a remarkable share considering that in the nation at large hydroelectricity accounts for only about six or seven percent. Because hydroelectric power is renewable, the Pacific Northwest more than other regions is in step with the rest of the world; hydroelectricity, most of it derived from dams, accounts for nearly ninety percent of the world's electricity drawn from renewable resources. Cheap, plentiful electric power has made the Pacific Northwest what it is today, and thanks to its renewable fuel, its great rivers, and to its dams, it would seem to be assured of a degree of continuity in its way of life, a cushion against the energy cutbacks that look to be in store for all of us.

Power is the fortune – critics might say, the misfortune – of the great River of the West. We probably would not build those same dams today, but we would still surely build dams. What started out as a love affair with dams has settled into a long-standing marriage, now of some seventy years, a familiar cohabitation most of us cannot imagine getting along without.

Dams are big, they stand out, they alter the course of nature. The damage they do to fish runs is easily pictured: they block rivers in a big way, so the fish obviously have a problem. For several decades, power and other benefits were a convincing argument for building big dams in America. If there had been no environmental movement, we might have gone on building them to the last site, and even this movement might have done no more that put up a temporary stumbling block. It took a single issue of that movement to make new starts on big dams in America all but unthinkable, at least for now, fish.

PART 4: DAMS, A SELECTION

I have selected most of the photographs of dams for this part of the book from the album Mac put together while working for the Corps of Engineers. They show a range of types of dams together with aspects of the work of building and operating them. I group the photographs by individual dams, and I introduce each group with a commentary. The first two groups are photographs of the first two dams on the Columbia River, most taken soon after they were completed, Bonneville and Grand Coulee. The next group is photographs of the Bureau of Reclamation's Shasta Dam in California. For his work in the Columbia Basin, Mac studied the spillway of this high, gravity-arch dam. The next group is photographs of the first dam Mac worked on, McNary. In addition to several photographs of parts of the dam on which he did the hydraulic design, there is a series of photographs of a temporary closure and raising of the river, an operation unprecedented in some ways. The next group is photographs of Lucky Peak Dam on the Boise River. This dam is not a run-of-the river, concrete, hydroelectric dam like Columbia and Snake River dams but a packed-earth dam fitted with a peculiar outlet works, built primarily for flood control and irrigation. Several of the photographs show tests conducted at the Bonneville Hydraulic Laboratory of models of structures to dissipate the excess energy of the high-velocity flow through the Lucky Peak outlet works. The next group is photographs of the construction and operation of the spillway of a dam built primarily for storage, towering Dworshak Dam on the North Fork of the Clearwater River. The final, largest group of photographs chronicles the stages of construction of the last American dam built on the Columbia River, John Day.

BONNEVILLE DAM

Various state and local bodies had long shown an interest in building a dam on the Columbia River at tidewater. The reasons they gave covered the bases: power, navigation, flood control, and irrigation, with different emphases. In 1916, the Oregon State Engineer worked out plans and estimates for a series of river projects, one of which was a dam at Bonneville, but nothing came of it as there was no market for the power. Portland General Electric Company made borings around the Bonneville site in 1929, but backed off because the cost was too great. Backed by the Corps' 308 Report's conclusion about the potential of the Columbia River, Roosevelt in the course of his first presidential campaign made a side trip to view the Bonneville site. He was also interested in the Grand Coulee site. Roosevelt's secretary of the interior initially rejected the Bonneville project on the grounds that the federal government could afford only one big dam in the Pacific Northwest, which was to be Grand Coulee. Bonneville Dam raised technical doubts too, about the suitability of the foundation rock. The Corps did more geological surveys and came up with a firmer site, the present Bonneville site, a few miles upstream from the one originally proposed at Warrendale. With that obstacle out of the way, special pleading by Oregon Senator Charles McNary and an Oregon representative brought Roosevelt around again, and in 1933 he authorized Bonneville Dam as a Public Works project. This would give the Corps its first experience in designing and building a dam that combined power and navigation in the same structure. (Wilson Dam had been designed by a private consultant.) It was decided that Bonneville would be a low, overflow gravity dam, and as such it did not break new ground in its basic design.

It was decidedly original, however, in certain of its major features, the most important of which were a new spillway design with structures for dissipating energy, a new concrete mix to minimize thermal cracking, and an extra powerful water turbine with adjustable blades. There was originality too in the scale of the construction; the coffer dams, navigation lock, gates, turbines, and fish facilities were all huge. The integration of the innovative work of separate branches of engineering – civil, chemical, mechanical, and electrical – was a signal achievement of the Corps, one which it would repeat at subsequent Columbia Basin dams.

There were two principal design problems, both of which the 308 Report had singled out: the dam had to accommodate a larger flood than did any other dam in the country, and it had to be kept from sliding on the soft, volcanic foundation. The solutions were, first, to make the spillway very wide and the spillway gates very large, fifty and sixty feet high and fifty feet wide. With the gates wide open, the spillway could pass a flood a third greater than the historic flood of 1894. As a further assurance against over-topping, the upstream channel was widened to lower the river at the approach to the dam. Second, stability of the dam was achieved by making the dam massive and wide and giving the base of the foundation a stepwise profile. Two related problems were the great variation in the volume of flow of the river, and the destructive energy of the water coming over the spillway. The first of these was solved by the big gates and also by a low sill, the second by two rows of baffles and a long apron below the spillway. In the hydraulic laboratory, the engineers studied scale models of the spillway gates and the river upstream from the dam, an important originality of another kind; for the first time the Corps modeled a spillway for a major dam. For all these reasons, Bonneville Dam held considerable engineering interest.

The powerhouse like the spillway had to be built to handle very large flows. It initially had six generators, to which four more were added later. The water turbines driving the generators were the Kaplan type, with adjustable blades to compensate for variable loads. Selected primarily because of its high efficiency under a wide range of loads, a Kaplan turbine could pass a volume of water each second large enough to fill an average three-bedroom house. The total capacity of the original powerhouse was just over a half million kilowatts, enough electric power to serve a city three times the size of Portland at the time, to some people a cogent reason to question the wisdom of building the dam.

The original plan called for a navigation lock to accommodate barges two abreast. Local interests pushed for a lock capable of passing an ocean-going ship, and on a visit to the dam under construction in 1934 Roosevelt expressed a desire for the bigger lock. The Corps of Engineers obliged; the lock that it built was capable of receiving an 8,000 ton vessel. The sixty-foot lift made it the highest single-lift lock in the world.

We come to the basic problem with Columbia Basin dams, fish runs. The starting point of the dam, the Corps' 308 Report, included fishways in its cost estimates. When Bonneville Dam was approved in 1933, the Corps after consulting with the US Bureau of Fisheries, Oregon and Washington fish and game agencies, and fishing associations assembled a group of fishery specialists. Opinion among them was divided between fish lifts or locks and fish ladders as the best way to preserve migratory fish runs; the final recommendation was for both, which was novel. The fishery agencies found the Corps cooperative, and a good deal more was spent on fish facilities than was first planned. In the end, Bonneville was fitted not only with lifts and ladders but also with a collection system and bypasses for fish, and a large Oregon fish hatchery near the dam was redesigned and relocated. When Bonneville Dam was built, nobody knew what effect it would have on the fish. The Washington State Game Commission predicted that ninety percent of the fish would be killed at Bonneville, even though the dam was fitted with fishways. The commission had little evidence, and its guess was overly pessimistic, but the optimists were wrong too.

The dam drew wide interest. Hundreds of thousands of visitors came to the site to watch it go up. Because of the volume of flow of the river and the depth of the foundation, the construction was monumental in scale, and at the same time it was done with extreme precision. The story of the building of Bonneville is one of boldness and ingenuity, but here and elsewhere in this book to do justice to the stages of construction would take us too far afield.

81. Bonneville Lock. Ca. 1940. Tug towing a raft of logs deep down in the lock. Navigation is one of the two main purposes of Bonneville Dam, electric power being the other. When the lock opened in 1938 it was the tallest in the world, but the record was short-lived; all of the subsequent locks built on the river were taller. Eventually, in 1993, a new lock was built at Bonneville. This photograph of the old lock was taken around 1940. Albert L. McCormmach.

82. Bonneville Lock. Ca. 1940. Same tug at river level. Albert L. McCormmach..

83. Bonneville Dam. Department of Ecology, State of Washington.

84. Bonneville Electric Generators. This is the interior of the second powerhouse, built in 1974-81, which doubled the dam's hydroelectric power capacity to 1,084,900 kilowatts. Bob Heines.

GRAND COULEE DAM

More than any other dam on the Columbia River, Grand Coulee Dam capitalizes on the local geology. During the last Ice Age, the Columbia River was blocked by ice at this site, creating glacial Lake Columbia. When to the east an ice dam containing a similar lake, Glacial Lake Missoula, collapsed, as it did repeatedly, floodwaters carved a new channel, Grand Coulee, into which Lake Columbia drained. When the ice melted, the river changed course again, leaving the canyon high and dry. The canyon, Grand Coulee, is enormous, measuring fifty miles in length, one to six miles in width, with walls rising 1300 feet or so. Built where the ice had once blocked the river, the dam reproduces the Ice Age in a controlled way. Part of the canyon is blocked off, forming a second reservoir, Banks Lake; water from the first reservoir behind the dam, Roosevelt Lake, is pumped into the second reservoir for storage, and from there it is released to flow by gravity over a large, arable plateau. Hydroelectric power generated at the dam works the pumps that feed a system of canals, siphons, and lakes extending more than a hundred miles south of the dam.

As far back as the late nineteenth century, suggestions had been made to build an irrigation dam at Grand Coulee. In 1904 the new Bureau of Reclamation considered a project in the vicinity, but decided against it on the basis of cost. Later, as a result of the Corps of Engineers' study of the site for its 308 Report, the Bureau recognized that hydroelectric power was a way to pay for an irrigation project, and it revived its interest in the site; in an independent report at the time, the Bureau supported the high dam recommended in the 308 Report. The Bureau's and Corps' plans went nowhere for a time. With the onset of the Great Depression, the Hoover administration was more concerned with balancing the budget than building big dams, and it was lukewarm about public power. Roosevelt, who succeeded Hoover, had a contrary opinion on public power, and he took a personal interest in Grand Coulee Dam.

The planners went to work again. They faced two major decisions: the type of dam, and the height, high or low. The Bureau of Reclamation considered a multiple-arch dam, but rejected it in favor of a more conventional, straight-axis, gravity dam like Bonneville. The height was more controversial. Local support was divided between a high and a low dam. Private power interests fearing unfair competition from the government wanted the dam to be as low as possible. Washington Senator Clarence Dill wanted a high dam as the centerpiece of a projected Columbia Basin Irrigation Project. Because of the expense, Roosevelt initially approved a low dam designed to produce power, though he stipulated that its foundation should allow for the height to be raised later at modest extra expense. The Bureau dutifully called for bids on a low dam with hydroelectric potential, but it too wanted a high dam for irrigation, and it planned accordingly. There were several arguments in favor of a high dam from the start: it was economical; pumps did not then exist powerful enough to lift water from the reservoir of a low dam to the coulee for storage and irrigation; a high dam would generate surplus electric power to subsidize irrigation; if a low dam were built first, it would be hard to make a good seal between it and a high-dam extension; and turbines designed for a low dam would not work efficiently for a high dam. The arguments would prevail, but not at first. Grand Coulee Dam was authorized as a PWA project in 1933, and construction started up on a low dam; Congress appropriated funding with that understanding. It was not until late 1934 that Roosevelt and the secretary of the interior were committed to the Bureau's high dam in its stead.

To realize the power potential of Grand Coulee, the electric generators ordered for it were the largest in the world, each with a capacity of over a hundred thousand kilowatts. There were originally to be two powerhouses, with a combined total of eighteen generators. When the first generator went on line in the fall of 1941, Congress had recently passed the Lend-Lease Act, resulting in a nation-wide increase in manufactures; the Bureau of Reclamation responded by speeding up its schedule for installing the remaining electric generators. Before the end of the year, Japan attacked Pearl Harbor, bringing America into World War II, and producing a yet greater demand for Grand Coulee power. By the spring of 1942 the next two generators were in operation, making Grand Coulee already one of the greatest power plants in the world; and five more generators, two of them borrowed from Shasta Dam, were installed before the war was over. The anticipated lull in demand for power after the war did not happen. Instead, the economy expanded, and the Pacific Northwest actually experienced a power shortage, something it had not done during the war; the Bureau of Reclamation again advanced its work schedule. In 1949 President Truman pressed a button at the White House to start up the first generator in the newly completed second powerhouse; and in 1951 the last generator was installed there. A third powerhouse at Grand Coulee became feasible after the Columbia River Treaty between the US and Canada, discussed below, went into effect in 1964; Congress authorized it two years later, and construction began the year following. Its generators, six in number, are huge, as is the power they produce; the diameter of their rotors is double the diameter of the rotors of the generators in the first two powerhouses and the rotors are three times as heavy. With the completion of its third powerhouse, Grand Coulee Dam was, for a time (again), the greatest producer of hydroelectric power in the world. There is also a smaller, fourth power plant at Grand Coulee for operating the dam and the storage pumps.

Grand Coulee is a multipurpose dam, though as usual the purposes are unequally served. The Corps of Engineers' 308 Report did not recognize navigation as one of the benefits of Grand Coulee; it was accepted that the dam would cut off commercial river transport to Canada. The report did recognize flood-control, although the reservoir behind Grand Coulee Dam was not intended to hold excess flow but to be kept full to maximize power output; however, after the 1948 flood, the reservoir was lowered for a period until upstream storage dams were built. The report stressed the other two main purposes of dams: power, discussed above, and irrigation, the Bureau of Reclamation's particular interest.

The Bureau made Grand Coulee Dam the center of its most ambitious planned agricultural settlement. We saw that the promoters anticipated an influx of settlers into a hardscrabble area where once there were only sagebrush, rattlesnakes, and a scattering of farmers and ranchers. To realize the full wealth of the Columbia River and the reclaimed land, the project was intended to support a large number of small family farms, and land speculation was to be ruled out. Owners were to be provided with ample, subsidized water and power to assure that their farms would prosper. This visionary goal was suspended in World War II, the dam then being used exclusively to power war industries.

While the war was still in progress, in 1943 Congress passed legislation creating the Columbia Basin Project. It allowed new settlers to buy farm lots of forty, sixty, and eighty acres up to 160 acres depending on the quality of the land. Owners of existing farms were permitted to keep 160 acres regardless of the quality of the land, but any acreage beyond that was to be sold at prewar prices. With this provision, the Bureau backed off from its original plan to restrict land holding to eighty acres, the start of a trend toward ever bigger farms; in the 1950s, when water from Grand Coulee began to fill irrigation ditches, a married couple was permitted to own 320 acres and to lease additional land, and in the 1980s the limit was increased again to 960 acres. Eventually the project would contain large agri-businesses and even corporate spreads with absentee landlords. The maximum number of persons living in the area of the project peaked at around 80,000 in the 1970s, and most of these were not the foreseen poor, Dust-Bowl refugees but financially comfortable Northwesterners. The original project was intended to irrigate a little over a million acres, but only two thirds of this acreage has been developed. The social goals that electricity was expected to achieve had been abandoned, Richard White writes, and instead, "electricity had become an end in itself." The paradise for a million new settlers, or for a scaled-down estimate of 300,000, to be accompanied by new industries and new cities, is a broken dream, but the power delivered by the dam has made the desert bloom through more productive farming, and with rural electrification – most of the rural population of the Northwest did not have electricity in 1933 when the dam was approved – it has brought comfortable living and labor-saving equipment to farmers, small and large.

We look at what Grand Coulee Dam meant for the fish. Before the dam was built, the Columbia River had the greatest population of salmon anywhere in the world, and most of the salmon migrated beyond the Grand Coulee site. The decision to build a fifty-story dam there obviously benefited irrigation, but it was the end of the line for the fish. It eliminated one thousand miles of streams and rivers available for spawning, mostly in Canada. The Canadian authorities had no problem with the dam, since their country had no commercial salmon fishery on the Columbia. Americans did have one, and for this and other reasons they took a measure of action, if belatedly. Midway through the construction of Grand Coulee Dam, it was decided to install temporary fish ladders, and it was also decided to trap and truck migrating salmon to their spawning sites above the dam. Also fish hatcheries were brought in to redirect the salmon that formerly passed the dam to tributaries of the Columbia below it. When no salmon were observed at the base of the dam two years after it was completed, the relocation program was judged a success. Now, decades later, no one can say just how well this experiment worked, since downstream dams have hopelessly complicated the history of the fish runs. The next dam downstream, Chief Joseph, is a permanent blockage of the river, having no fishways. Priest Rapids Dam far down the river is the end of the last important fall chinook salmon spawning grounds on the main-stem Columbia River.

Grand Coulee Dam probably can never be considered complete. Economic and environmental developments that affect it are unpredictable; jurisdictions at multiple levels, international, national, state, and local, all have their say; and its operation is entangled with a good number of often-conflicting issues: government spending, hydroelectric power, flood control, irrigation and farm policy, reclamation and conservation, recreation, Indian rights, environmental damage, wild life, and today the most politicized of them all, fisheries.

85. Grand Coulee Dam. Panoramic frontal view of the dam after the third powerhouse had been added, on the left. GFDL.

86. Grand Coulee Dam and Powerhouse. When this photograph was taken, in the 1970s, it was the last time for several years that the dam would spill water. Albert L. McCormmach.

87. Grand Coulee Electric Generators. This photograph shows the interior of one of the original two powerhouses. Each powerhouse was fitted with nine generators, each rated at 125,000 kilowatts. Originally, before the stators were rewound, they were rated at 108,000 kilowatts. US Bureau of Reclamation.

88. Grand Coulee Water Turbine. This Francis turbine, rated at nearly 1,000,000 horseower, is intended for one of the six generators stalled in the dams third powerhouse. Originally, they were rated at 700,000 and 600,000 kilowatts. Wikipedia Foundation.

SHASTA DAM

Like Grand Coulee Dam, Shasta Dam on the Sacramento River is a New Deal, public works dam built by the Bureau of Reclamation. Shasta is the northern half of a two-dam, major storage project in California's 500-mile-long Central Valley; the southern half is Friant Dam on the San Joaquin River. (The entire Central Valley Project contains a total of twenty dams and eleven power plants today.) The northern sector of the valley receives considerable rainfall; the southern sector is much drier, and the San Joaquin River accordingly has a much lower flow than the Sacramento. The goal of the project is to store excess water at Shasta Dam and channel it to the south. Built as a high dam, Shasta produces a modest quantity of hydroelectricity, the main purpose of which is to work the pumps that supply water to the upper San Joaquin Valley. The dam also provides flood control, navigation, and municipal and industrial water. The Central Valley Project is multipurpose and immense, in both respects bearing comparison with the Columbia Basin Project at Grand Coulee on the Columbia River and the Boulder Canyon Project on the Colorado River.

89. Shasta Dam. A gravity-arch, concrete dam similar to the Bureau's earlier Hoover Dam. Like that dam, Shasta was regarded as a great engineering feat at the time. The powerhouse is on the lower left. US Bureau of Reclamation.

Preceded by a similar state initiative, the Central Valley Project was authorized in 1937; the Bureau of Reclamation began construction of Shasta Dam the next year, completing it in 1945. Earlier, California state engineers had proposed a 500-foot-high dam for the site; the federal engineers of the Bureau proposed a somewhat higher dam, 560 feet, the height of the Bureau's Grand Coulee Dam; the height of the constructed dam is higher yet, 602 feet. Like Grand Coulee, Shasta is a massive, solid-concrete, gravity dam. Other designs for it were considered – thin-arch, multiple-arch, earthfill, and rockfill – but they had not been tried on a very high dam like Shasta. Cost was a consideration too; it was then cheaper to build a spillway and outlet works for a dam made of concrete than for a dam made of other materials. Shasta Dam differs from Grand Coulee in appearance, curved instead of straight, but in the design of the dam no consideration was given to the extra strength provided by the arch; gravity alone resists the water load. Shasta is the second largest concrete dam in the United States after Grand Coulee. In hydroelectric power it is not in Grand Coulee's class; rather it compares with Ice Harbor Dam on the Snake. Its reservoir, with half the storage capacity of Grand Coulee's, 4.5 million acre-feet, is the largest in California.

90. Shasta Dam. Five fifteen-foot-diameter penstocks direct water from the pool behind the dam to the turbines in the powerhouse below it. Albert L. McCormmach.

91. Shasta Dam. Spillway and penstocks. The spillway is often described as the highest man-made waterfall in the world. It has three drum gates at the top, each 28 feet high and 110 feet long, and in addition it has eighteen 8 1/2 foot diameter outlets on the face of the dam at three levels for releasing water at different pool elevations. In this photograph, it is releasing water from the lowest outlets. Like certain dams in the Columbia Basin, Shasta Dam blocks the fish migration and the passage of cool water, both causing a decline in fish population. To help the fish down river, cool water can be released from lower outlets on the spillway, though at the cost of water lost to power. A temperature control device is now installed on the upstream face of the dam, which selects water from different depths, at different temperatures, feeding it into the penstocks, with no loss to power; this type of device is discussed in the section on Dworshak Dam in the Columbia Basin. Albert L. McCormmach.

92. Shasta Dam. Discharged from an outlet in the face of the dam, the stream is moving as it should, tangent to the spillway. To dissipate its energy, the stream falls about 150 feet into flip buckets at the toe of the dam. Albert L. McCormmach.

93. Shasta Dam and Powerhouse. This view of the dam shows the powerhouse in the foreground. It has five generator units with a total hydroelectric capacity of 676,000 kilowatts. Like Bonneville and Grand Coulee, Shasta Dam was important in World War II, supplying power to shipyards and aircraft factories. In recent years, the Bureau of Reclamation has proposed raising the height of the dam for greater storage, and it would install a new powerhouse, increasing the output to that of a big dam on the Columbia River such as The Dalles. Albert L. McCormmach.

MCNARY DAM

Umatilla Rapids on the Columbia River were a formidable obstacle. The river flowed thirty-five miles an hour past them, and they were so shallow a man could stand in them. At first stern wheelers and then tugs and barges broke up in these rapids, and men drowned. River transport and farming interests had long pressed for a dam at the rapids, and for twenty-five years the rapids had been studied as a potential site for development by the Corps of Engineers and by the Bureau of Reclamation. In 1922 the Corps recommended a dam at the rapids as one of a series of dams on the Columbia, all under fifty feet, to facilitate river passage from the ocean to Idaho. At about the same time the Bureau of Reclamation recommended a dam as the principal facility for an irrigation project. In 1933, the Corps proposed a different project, a power dam in addition to a navigation lock, but power subsidies were not in favor then and anyway power was abundant and no action was taken. Again in 1937, the Corps recommended a dam for power and navigation, though at the time it recognized that the demand for power was not yet sufficient to pay for a dam. During World War II a House committee approved a Umatilla dam for national security among other purposes. When Congress finally authorized Umatilla Rapids Dam in 1945, it moved quickly on the funding, since much preliminary planning of the dam had already been done during the war. The dam was renamed McNary, after the Oregon senator mentioned earlier for his role in persuading Roosevelt to authorize Bonneville Dam.

We see the established methods of dam construction at work at McNary Dam. The first cofferdam proceeded from the north shore in a U-shape, enclosing most of the spillway, fishways, and the navigation lock. The second cofferdam was a minor dam proceeding from the south shore, enclosing a portion of the powerhouse and a fish ladder. The third cofferdam proceeded from the south shore, enclosing the rest of the spillway and the powerhouse. To close the third cofferdam, it was necessary to divert the river from its natural rapids to flow through a portion of the spillway on the north shore, which purposely had not yet been built to its full height. The procedure was unusual and its execution dramatic, an example of how in the application of standard methods problems came up that called for innovative solutions.

The closure concerned a gap of 240 feet between the north and south cofferdams. The entire river poured through the gap, making it impossible to enclose with a final cofferdam. The object of the closure was to raise the river sufficiently for it to flow through the unfinished spillway bays, enabling a cofferdam to be built at the gap. Within the Corps, there was a sharp difference of opinion on how to do this. Steel cells, timber cribs, and a scuttled Liberty ship were among the proposals. The method the Corps settled on was suggested by its Passamaquoddy Tidal Power Project in Maine, a New Deal undertaking to produce hydroelectric power from the rise and fall of the tides, and also to create jobs during the Great Depression. The method used at McNary was to simulate an act of God, analogous to a piece of mountain falling into the river: it was to fill the gap with artificial stones in the form of pre-cast concrete tetrahedrons, each nine feet high and weighing twelve tons. The tetrahedrons were transported by a cableway supported by movable towers in the river and carefully dropped at designated locations, where because of their sharp corners they remained, locked in place, instead of rolling and tumbling downstream. They behaved as predicted by model studies, as was determined when that section of the riverbed was dried by the cofferdam. It took three thousand tetrahedrons to close the gap in an operation that lasted six weeks. The closure was the most difficult ever tried, and people came from all over the world to see it. Photographs of the closure give you a feeling for the power of moving water and for the brute force required to control it.

During the construction of McNary Dam, a number of temporary structures were built to keep the fish moving, supplemented by temporary fish hatcheries and other expedients: fish ladders, navigation lock, a huge round steel bucket lifted by a derrick, and an even bigger fishnet moved by a mobile crane, and at one point old-fashioned nets handled by experienced Indian fishermen. Permanent fish ladders were built at both shores. At that time the only federal dam on the Columbia River with fish ladders was the first, Bonneville, and in the ten years since that dam had been completed, no decline in the upstream fish runs had been credited to it. The fish ladders at McNary were confidently modeled after those at Bonneville, an evident success.

The reservoir behind McNary Dam rose above Indian burial grounds, a problem repeatedly encountered with dams in the Columbia Basin. After consultation with tribal chiefs, the decision was made to leave the grounds undisturbed under water. Before the reservoir was filled, grave robbers went in. The Corps received severe criticism from the Indians for allowing this wanton destruction, and rightfully so. Eight communities along the shore were affected by the reservoir, three of them completely inundated, and a system of levies had to be built to protect Tri-Cities. The dam-builders moved in like a geological force.

Ports were built and industrial enterprises were set up around the shore. The upper end of the reservoir marked the end of Lower Columbia River navigation – navigation was picked up again above Grand Coulee Dam – but because the reservoir backed up to the mouth of the Snake River, navigation was extended along that course. Eventually, with the aid of eight locks on eight federal dams, four of them on the Lower Snake River, slack-water navigation enabled barges to go up the Columbia River as far as Tri-Cities, 325 miles, and 144 miles up the Snake River as far as Lewiston, Idaho. McNary Dam was an Inland Empire transport resource as well as power plant.

McNary Lock and Dam took seven years to build. The dam is nearly a mile and a half wide, with a hydraulic height of ninety-two feet, nearly the height of a ten-story building, the same as the hydraulic height of later federal dams on the Columbia and Snake Rivers. At the time it was built, the navigation lock was the highest single-lift lock in the world, though the dams at Ice Harbor and John Day would have higher ones. Its twenty-two spillway bays are capped by vertical lift gates, roughly the same size as the tainter gates on most of the Corps' dams, around fifty feet square, and operated by cranes. The powerhouse delivers close to a million kilowatts. When President Eisenhower gave the main address at the dedication of McNary dam, thirty or forty thousand visitors turned out. In 1954, a new big dam was a big deal.

94. McNary Dam. In this aerial view of the completed dam from the Oregon side, we see the earth embankment and powerhouse in the foreground and the concrete spillway, lock, and a fish ladder in the background. US Army Corps of Engineers.

95. NcNary Dam. Piers and the beginning of the "S" - shaped spillway. US Army Corps of Engineers.

96. NcNary Dam. Columbia River closure, 1950. Midway through the construction of the dam, it was necessary to close the gap between the north and south cofferdams, raising the river and redirecting its flow. This aerial view shows the operation in progress. US Army Corps of Engineers.

97. McNary Dam. Columbia River closure, 1950. Frozen gravity. The twelve-ton concrete tetrahedron is caught in mid air by this fast-speed photograph as it is dropped in place. US Army Corps of Engineers.

98. McNary Dam. We see the cableway that drops the tetrahedrons, positioned at a slant across the gap. US Army Corps of Engineers.

99. McNary Dam. Columbia River closure, 1950. Progress. US Army Corps of Engineers.

100. McNary Dam. Columbia River closure, 1950. The barrier of tetrahedrons is about to break surface. US Army Corps of Engineers.

101. McNary Dam. Columbia River closure, 1950. The first tetrahedrons surface, and the river is starting to rise. US Army Corps of Engineers.

102. McNary Dam. Columbia River closure, 1950. Looking north, the completed closure. At the upper left, we see the river passing through the low bays of the partially completed spillway, the immediate goal of the closure. US Army Corps of Engineers.

103. McNary Dam. Rhythmic pattern of piers, shadows, and flows. Albert L. McCormmach.

104. McNary Dam. Stilling basin with two rows of baffles. Albert L. McCormmach.

105. McNary Dam. Downstream closeup of the river flowing through low spillway bays, with only the piers visible. Albert L. McCormmach.

106. McNary Electric Generators. Interior of the powerhouse showing the fourteen 70,000 kilowatt generators, with a total capacity of 980,000 kilowatts. The power plant is unique in that it has two additional units to power the dam itself, making it self-sustaining and a backup in case the entire Northwest grid should break down. A second powerhouse was authorized in 1986, but it was canceled five years later. US Army Corps of Engineers.

LUCKY PEAK DAM

When the Corps of Engineers and the Bureau of Reclamation made their peace in the Pacific Northwest in the late 1940s, they agreed to leave in place any projects already lined up. That included Lucky Peak Dam on the Boise River, a tributary of the Snake River, the only multipurpose dam built by the Corps in the Bureau's territory.

Lucky Peak Dam was a flood-control dam at the start, a natural project for the Corps. It came about in an altogether typical way. The lower Boise River was given to flooding, and following a big flood in 1943 a water conservation lobby appealed to its Washington delegation. In order to win support for a dam, there needed to be an additional use, which was obvious, irrigation. After its approval by a subcommittee, Congress authorized Lucky Peak Dam in its Flood Control Act of 1946; it allocated planning funds two years later, and construction funds the year after that. Lucky Peak Dam was built between 1949 and 1955.

The site is downstream from two impressive Bureau of Reclamation dams. The earliest of them, Arrowrock Dam, at 350 feet was the highest concrete dam in the world at the time it was built, in 1915; it is a gravity-arch dam, built mainly for irrigation. The other Bureau dam, Anderson Ranch, is a multipurpose, earthfill dam, which in turn was the highest dam of its kind in the world for a time, at 456 feet. Begun in 1941, Anderson Ranch Dam was still under construction when work on Lucky Peak started up; it was completed before Lucky Peak, in 1950. Like Anderson Ranch, Lucky Peak is an earthfill dam with a large reservoir, comparable to the upstream reservoirs. At 340 feet Lucky Peak Dam is not as high as Arrowrock and Anderson Ranch, but it is remarkable in a different way, in the original giant arc of its discharge.

Lucky Peak Dam was a project of the Portland District. Mac left the district in 1948, and since the design of the dam was only decided in 1949, it is unlikely that he had a part in it, but later when problems developed he was brought in. As an earthfill dam, Lucky Peak Dam looks different from the concrete dams on the Columbia and Snake Rivers. The spillway is not on the face of the dam but is a separate structure, a long concrete section on the left abutment, rarely used. The original design called for a single outlet tunnel, twenty-three feet in diameter and 1,200 feet long, embedded in the same abutment as the spillway, and terminating in a structure called a "manifold," containing six five-by-ten-foot gates. During construction, the same tunnel was used to divert the river. In the normal operation of the dam, the river flowed through the tunnel into the manifold, where the six open gates directed it onto six upward-slanted concrete blocks, flip buckets. The high-velocity flow, up to seventy miles an hour, coming off the flip buckets formed a rooster-tail plume, the jet traveling for some distance before splashing down. The purpose of this strange-looking outlet works was to dissipate the very considerable energy of the flow, minimizing erosion of the stilling basin. It was in connection with the outlet works sustaining these high velocities of flow that problems came up, the occasion for Mac to take the photographs included in this part of the book.

The single outlet tunnel was unsatisfactory in a fairly obvious way. If it had to be shut down for repairs in some part of the system, there would be no river downstream, only an empty stream-bed. The Boise water supply and sewage treatment would stop in the event, and fish would be stranded and die. The city made the reasonable case for a second outlet to guarantee it would not run out of water. The Corps agreed with the city, and in 1976 Congress authorized a second outlet tunnel. Hydroelectric power was not one of the original multiple uses of the new dam, a rare omission for a federal dam in the Pacific Northwest. Energy, as it happened, was much on the American mind in the 1970s, a decade of energy crises, and it was evident from the rooster tail at Lucky Peak that substantial energy went to waste there. A principle reason the Corps wanted a second tunnel was that by diverting the river through it, it would enable the first tunnel to be fitted with electrical generators. Following a recommendation to this effect in 1978, the Corps entered into an agreement that allowed the irrigation people, the Boise Project Board of Control, to build a power plant and to market the power it delivered.

Nearly everyone got something from Lucky Peak Dam after the addition of a second tunnel. The only losers were the visitors, who no longer could count on seeing the rooster tail, a popular attraction. The lobbyists got the added flood protection they wanted, 300,000 acre-feet of storage. This together with the reservoirs at Arrowrock and Anderson Ranch Dams provided around a million acre-feet of storage, a considerable, though not absolute, protection for Boise from flooding. The irrigators got the stored water from the dam for distribution over a large swath of arid land in Idaho and Oregon, and they also got the revenue from sales of hydroelectric power. The city of Boise got a new sewage treatment plant from the Environmental Protection Agency. The Fish and Game Department was assured a minimum flow for the fish. And the Corps got a chance to demonstrate its new skills in making its plans in open forums. The Corps and the Bureau together with the owner of the powerhouse, the Boise Project Board of Control, jointly operate the reservoirs of the dams on the Boise River in what is recognized as a model of inter-agency cooperation. Lucky Peak after its conversion to a full multipurpose dam is a civic lesson in how to achieve harmony among competing interests on a river.

107. Lucky Peak Dam. Unlike the spillways of the concrete dams shown in this book, the Lucky Peak spillway is not built into the dam, but instead there is a 600-foot-long, free-overflow, concrete spillway located on the left abutment; it is to the right in the photograph. The dam also has an outlet works on the left abutment. A powerhouse was added there in 1984-86; it too is shown in this up-to-date photograph. The use of the face of the dam as a billboard for a slogan, Keep Your Forest Green, however worthy the thought, might seem tacky to some. US Army Corps of Engineers.

108. Lucky Peak Dam. This photograph shows steep, barren land forms at the site on the Boise River in Idaho. Albert L. McCormmach.

109. Lucky Peak Dam. Construction underway on an outlet tunnel. Albert L. McCormmach.

110. Lucky Peak Dam. 1959. Discharge through gates of the outlet works, producing the much photographed rooster tail. Albert L. McCormmach.

111. Lucky Peak Dam. 1959. Mac who posed for this photograph had come to observe friction tests on the outlet works under conditions of high discharge, here 4,540 cubic feet of water per second. Albert L. McCormmach.

112. Lucky Peak Dam. 1959. Flip buckets, described in the next caption, reduce erosion downstream, but the jet coming off them is turbulent and spreading, entraining air and producing a heavy spray that can can cause damage to structures and the environment. Flip bucket 3 was cut down to reduce spray, a problem at Lucky Peak. Albert L. McCormmach.

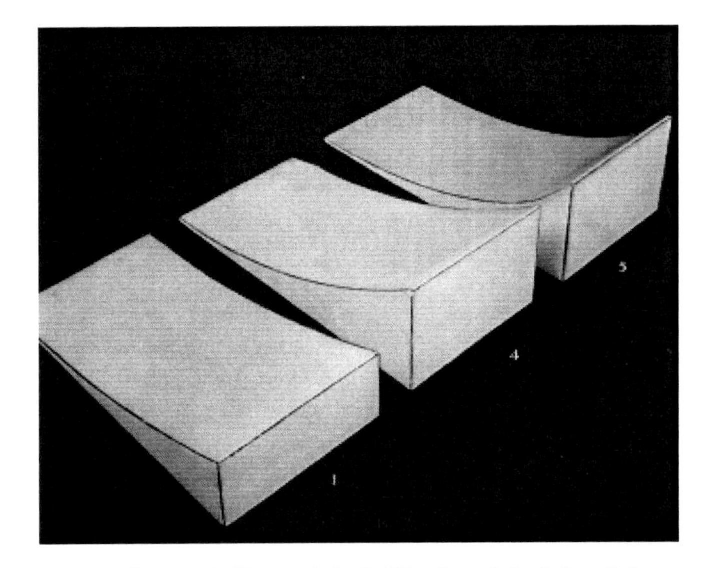

113. Lucky Peak Dam, ModelTesting. Models of three differently shaped flip buckets for Lucky Peak Dam were tested at the Bonneville Hydraulic Laboratory. Flip buckets are large, shaped-concrete blocks designed to abruptly lift high-velocity discharge, dissipating some of its destructive energy. Some energy is also dissipated by friction of the flow through the flip buckets. US Army Corps of Engineers.

114. Lucky Peak Dam, Model Testing. To the far left is the "manifold," which receives the flow from the outlet tube of the dam and controls its release downstream by six gates. In front of each gate is a flip bucket. These tests were part of a model study conducted at the Bonneville Hydraulic Laboratory of the outlet works and stilling basin at Lucky Peak. The goal was to design flip buckets for dissipating energy and a manifold that was free of cavitation. US Army Corps of Engineers.

115. Lucky Peak Dam, Model Testing. Trial of flip buckets, all six gates open. US Army Corps of Engineers.

116. Lucky Peak Dam, Model Testing. Model flip buckets in action, viewed from downstream. Operators observe patterns of flow in the basin using bits of light material. US Army Corps of Engineers.

DWORSHAK DAM

The 1948 revision of the Corps' 308 Report proposed a big storage project in the basin of Clearwater River, another tributary of the Snake River, and as an alternative it suggested building two dams on the North Fork of the river, one of these at Bruce's Eddy. This became the site of Dworshak Dam. A storage dam, which interrupts the natural flow of a river, making it more useful to humans, is a high dam with a big reservoir, in contrast to a run-of-the river dam. There is only a handful of storage dams in the Columbia Basin, but they play a major part in river control: by storing excess water during high flow and releasing it during low flow, they contain floods and increase hydroelectric power output. They are the key to the operation of the dams as a system.

Columbia Basin storage dams are a diverse lot: three public-utility and private-power dams, two Bureau of Reclamation multipurpose dams, three Corps of Engineer multipurpose dams, and three Canadian dams. The Bureau's dams are Grand Coulee and Hungry Horse, a 564-foot gravity-arch dam on the South Fork of the Flathead River in Montana, which along with Dworshak Dam was recommended in the 1948 Plan. The Corps' dams are Dworshak

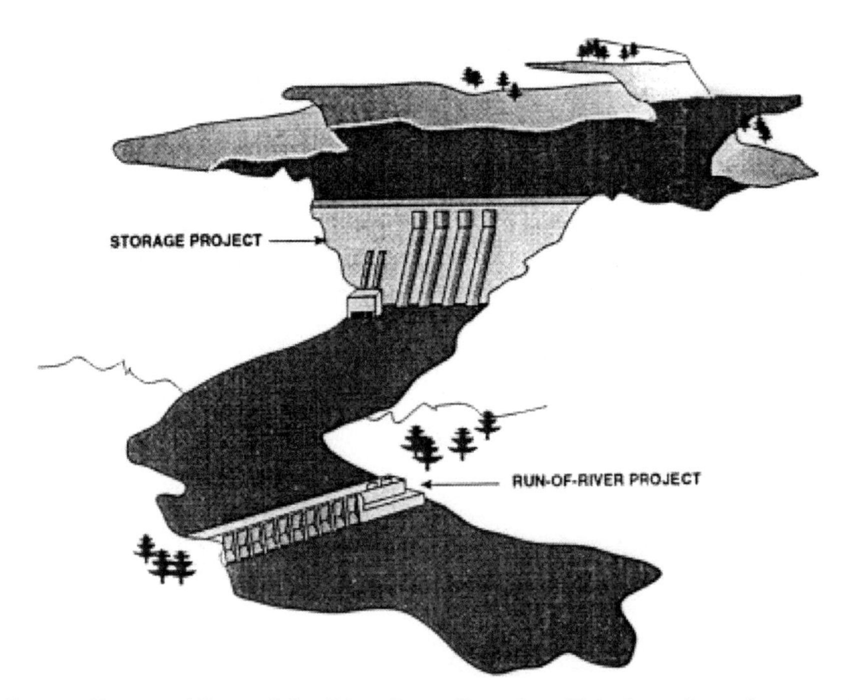

117. Storage Dam and Run-of-the-River Dam, Drawing. This shows how the operations of the two kinds of dams are coordinated. Water released from the upstream high dam provides the downstream low dam with a uniform flow for the production of power. US Army Corps of Engineers.

Dam, the 442-foot gravity dam on the Kootenai River in Montana, Libby Dam, and small Albeni Falls Dam on the Pend Oreille River.

The Canadian storage dams are important, since Canada receives thirty percent of the Columbia Basin runoff. Power shortages after World War II together with a need for flood-control prompted the US and Canada to sign the Columbia River Treaty in 1961; Canada ratified it in 1964. The treaty called for the two countries to build four dams primarily for storage: in addition to Libby Dam, they are Mica, Duncan, and Hugh Keenleyside Dams in British Columbia. Through the operation of its dams, which account for about half of the total storage capacity of the Columbia Basin, Canada insures that downriver dams in the United States meet their hydroelectric needs regardless of seasonal variation in river flow. In exchange for its cooperation, Canada receives a half share of the revenues from the extra power generated at American dams. Only one of Canada's three storage dams, Mica, also develops hydroelectric power, but together they make possible a fourth Canadian dam in the Columbia Basin, the run-of-the-river Revelstoke Dam, each of the two dams developing nearly as much hydroelectric power as John Day Dam in the United States, itself second only to Grand Coulee in capacity.

Like the Corps' other storage dam Libby, Dworshak Dam controls floods,

118. Dworshak Dam. US Army Corps of Engineers.

119. Libby Dam. Completed in 1975, in accord with the Columbia River Treaty with Canada, this Army Corps of Engineers storage dam is located on the Kootenai River in Montana, the third largest tributary of the Columbia River after the Snake and the Clark Fork-Pend Oreille. Its ninety-mile-long reservoir holds almost six million acre-feet of water. US Army Corps of Engineers.

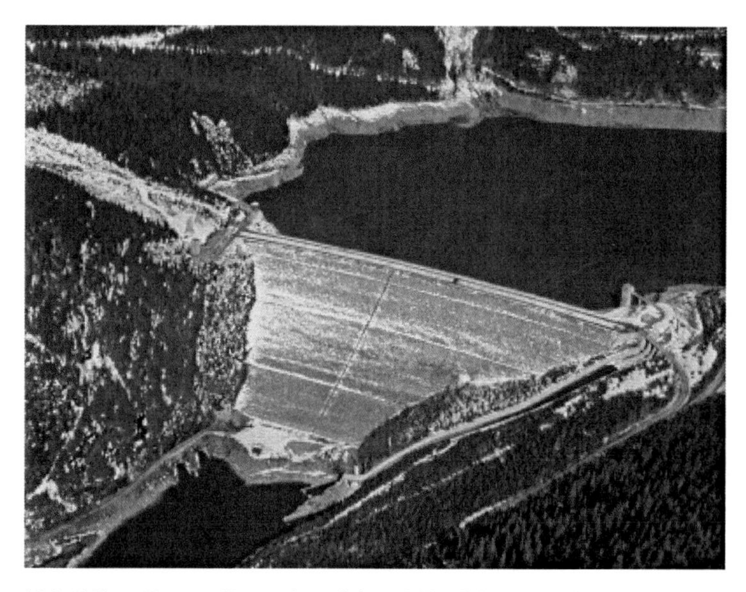

120. Mica Dam. Completed in 1973, this earthfill dam is the northernmost dam on the Columbia River, in British Columbia. Like Libby a Columbia River Treaty dam, its reservoir holds twice as much water as Libby's, twelve million acre-feet. This is the only Treaty storage dam that is also a major hydroelectric dam, with a capacity of 1,736,000 kilowatts. Wikimedia Foundation.

121. Revelstoke Dam. This is a concrete gravity dam on the Columbia River below Mica Dam in British Columbia. Its reservoir backs up about a hundred miles to Mica Dam, but since most of its flow is water released from the Mica reservoir, it operates as a run-of-the-river dam. In generating capacity, it is comparable to Mica Dam, 1,840,000 kilowatts. Wikimedia Foundation.

and it also generates power at the site and provides for upstream navigation and river transport of logs; that is, it is a multipurpose dam, though its uses do not include irrigation, as Bureau dams do. Mac was involved with this dam from the beginning, as we see from the photographs he took of Bruce's Eddy before there was any construction.

Bruce's Eddy – a "V" bend in the river with an eddy at the base of the V, where an engineer named Bruce lost his life at the turn of the century – lies immediately above a deep granite canyon, an ideal location for a dam. (The dam was called "Bruce's Eddy" at first, but like McNary Dam, it was renamed after a senator from the state.) As usual, the site had a history, having been studied by power and light companies in the 1920s and 50s. The federal government's plans for developing it date from a Middle Snake Report in 1953. Action was delayed for several years while negotiations with Canada over Columbia water storage were underway and while local opposition from fish and game people was weighed. Congress approved planning funds in 1958. Just at this time the region's projected power needs went up, and as a result the scope of the project was revised upward. The original plan was for a dam 570 feet high; the revised plan was for 717 feet. This height, being almost without precedent, provoked a heated disagreement over the kind of dam to build. There were advocates of a thin-arch dam, a gravity-arch dam, a rockfill dam, and the one that won out, the standard Corps concrete gravity dam. The main arguments for the latter were safety and cost, in that order. Construction began in 1966, and the dam began operation in 1972.

The dam has unusual features due to its great height. The face of the dam resembles Shasta's with its very tall, gated-spillway, only the spillway at Shasta is centered and the one at Dworshak is to the side. Dworshak Dam spills water from two elevations: there is a pair of bays with tainter gates placed high up the dam for drawing down a full reservoir, and there are three outlets placed far down the face of the dam for use when the reservoir is low. There is a steelhead migration on the river, and because fish passage is unfeasible in a dam so tall, the Corps built a steelhead hatchery, the world's largest, downstream as the alternative. Fish are also a major reason for conspicuous, columnar intake structures on the upstream face of the dam, which control the temperature of the river downstream from the dam. (The Bureau added a similar device to Shasta Dam in the 1990s.) The temperature in an unobstructed river is nearly the same at all depths, but the temperature in a reservoir is colder near the bottom where little sunlight penetrates, which is where water is taken from for release downstream, and this water has a different oxygen content. It was found from computer models of the operation of Dworshak Dam that the abnormally low temperature downstream would be unacceptable to fish. The solution was to build columns with three tiers of gates to take in water from three levels of the reservoir with three temperatures and three concentrations of oxygen, the selected water then being directed into the powerhouse penstocks.

The Corps awarded the biggest contract in its history to the builders of this dam. The dam called for over half the quantity of concrete that went into Grand Coulee Dam, the standard of bigness in North American concrete dam construction for all time. When it was completed it was, and it still is, the third highest dam in the United States, behind Oroville Dam in California, the highest, and Hoover Dam. Of its type, the Corps' preferred straight-axis, solid-concrete, gravity dam, it was the highest and biggest in the Western Hemisphere. The high dam made possible a large storage reservoir, with a capacity of around three and a half million acre-feet, three quarters the capacity of Shasta's reservoir. Not by its hydroelectric output, which was inferior to that of Columbia River dams, but by these other measures – mass, height, and storage – Dworshak is the biggest dam in the Walla Walla District of the Corps of Engineers.

Much of the technical interest of this dam lies in its construction. Its location, one of the colder parts of America, required great care in the temperature control of the concrete. Because of the massive size of the dam, engineers and contractors looked for ways to speed up construction. One of their innovations was to expedite the production of aggregate from local granite for concrete. Normally, the granite would be quarried and hauled by truck to the place where it would be crushed. Instead, at Dworshak a tunnel was run under the quarry ending in a chamber, inside of which a great crushing machine was installed. A hole was drilled from the quarry above into the chamber, and the granite was dumped down into the crusher, which broke up four-foot boulders into six-inch pieces at the rate of 2,000 tons an hour. The crushed granite was then carried by a belt running through the tunnel to an aggregate plant, which was capable of producing 1,400 tons of aggregate and sand an hour for delivery to a concrete batching plant. From there the concrete was carried in huge buckets moving at twenty-five miles per hour along three cable-ways spanning the river to the sites on the dam where it was laid.

Problems that arose in the finished dam held a technical interest of their own. Despite the caution with which the concrete was laid, the great temperature variations at the site caused major thermal cracks to develop in the monoliths, resulting in leakage, requiring a large number of drains to be drilled to relieve the pressure, followed by sealing. The very high velocities of flow prevailing at the dam presented more serious problems. In anticipation of these, and because of the uncommon use of the chute for both spillway and outlet works, very extensive tests were made at the Bonneville Hydraulic Laboratory on two scale models, one for the spillway, stilling basin, and tail bay, and the other for the right-hand outlet conduit and regulating valve. The dam itself was regarded as a kind of laboratory model, a prototype to be fitted out with instruments, which over the course of several years' operation were intended to furnish engineers with hydraulic information badly needed for the construction of any other very high dams. To this end, each of the three outlet conduits was given a different surface, as well as being made of superior concrete; and the concrete of the spillway and related structures was faced with a layer of especially dense, high-strength concrete two or more feet in thickness. Guided by the hydraulic laboratory tests, hydraulic and structural designers were able to forestall some but not all of the problems. The outlets developed serious cavitation erosion, removing concrete to a depth of six inches or more, a problem which was corrected by patching and resurfacing with fiber-reinforced concrete; and as happened at the Corps' other high storage dam, Libby, the stilling basin suffered serious abrasion erosion caused by rocks and gravel circulating over the surface of the concrete, with massive removal of concrete and steel up to several feet in depth. The sill at the end of the apron dissipates energy, but it also traps debris, and at the dam a structure was added downstream from the sill to prevent eddies from depositing debris on the apron; damage to the stilling basin was repaired by replacing the lost concrete and giving it a stronger surface. Dworshak Dam did indeed prove to be an engineering laboratory.

122. Dworshak Dam. Bruce's Eddy, 1960. Mac took many photographs of this bend in the North Fork of the Clear Water River in Idaho. We are looking upstream from the axis of the proposed Bruce's Eddy Dam, later renamed Dworshak Dam. Albert L. McCormmach.

123. Dworshak Dam. Construction, early stage. Albert L. McCormmach.

124. Dworshak Dam. Construction of spillway and outlet chute, viewed from the top. Albert L. McCormmach.

125. Dworshak Dam. Construction of spillway and outlet chute, viewed from the bottom. Albert L. McCormmach.

126. Dworshak Dam. Lower part of the completed spillway. Albert L. McCormmach.

127. Dworshak Dam. Dworsak has two spillway bays controlled by tainter gates at the crest, typically large, 50 by 56 feet. At a lower level, there are three regulatory outlets running through the dam, which are used when the stream exceeds the capacity of the powerhouse and the reservoir is lower than the spillway crest and also when cooler water is required for the safety of downstream fish. The outlets at the entrance measure twenty-two by sixteen feet, tapering to twelve by nine feet at the tainter gates inside the dam. Devices such as strobe lights are used to discourage fish from entering the outlets. In this photograph, we see all three outlets discharging. Albert L. McCormmach.

128. Dworshak Dam. Another view of the same operation, showing the placement of the outlets with respect to the spillway, the face of the dam, and the powerhouse. Albert L. McCormmach

129. Dworshak Dam. This late-in-the-day photograph shows the dam in its setting under familiar conditions, snow on the forested hills with low fog over the crest. We see discharge from one of the two spillway bays as well as from all three regulatory outlets. Albert L. McCormmach.

130. Dworshak Powerhouse. Powerhouse nearing completion, seen from the top of the dam. Two twelve-foot and four nineteen-foot penstocks pass downward through the dam to the powerhouse at the toe, which holds three generator units, with a total capacity of 400,000 kilowatts. Three more units had been planned, bringing the total to 1,160, 000 kilowatts, the capacity of a major power dam. But it was determined that this would cause downstream damage, and they were cancelled. Albert L. McCormmach.

JOHN DAY DAM

A dam at the John Day Rapids was first seriously considered in the Corps' 308 Report, and it was included in the updating of that report in 1948, which made it an integral part of the development of the Columbia Basin. Congress approved the dam in 1950, appropriating the first funds in 1956; construction began in 1958 and finished in 1971. The Corps considered a number of criteria in selecting the site, not all of which could be simultaneously satisfied. The main conflict was between the recommendation to locate the dam below the entrance of the John Day River to maximize flood control, and the recommendation to locate it above the entrance to reduce the number of dams that John Day salmon and steelhead have to pass on their upstream migration to their spawning grounds. The fishery groups, and the fish, lost this one.

Like McNary, John Day Dam required three stages of cofferdams, each enclosing a section of the eventual permanent dam. Again, the first cofferdam proceeded from the north shore, extending a good ways out into the river. Concrete was poured there for the navigate lock, the north-shore fish ladder, and twenty-two spillways. The second cofferdam was built from the south shore, enclosing the powerhouse and the south-shore fish ladder. The third cofferdam allowed for the completion of the spillway. In building Columbia River dams, the Corps had found its rhythm.

131. John Day Dam. This Army Corps of Engineers run-of-the-river, concrete gravity dam on the Columbia River, just below the mouth of the John Day River, is around a mile and a half long. In this upstream view, the huge powerhouse is on the right. US Army Corps of Engineers.

The finished John Day Dam is larger than McNary Dam, about the same size as The Dalles Dam. Innovative techniques were used in the cofferdam stages, calling for steel cells that were among the largest ever built, over eighty feet high. The hydroelectric capacity of John Day Dam is, as mentioned, second only to Grand Coulee's, and the navigation lock is the highest single-lift lock in the US, at 113 feet. The model of the dam at Bonneville Hydraulic Laboratory was itself big, longer than two football fields. Even the planning and construction of John Day Dam took longer than most, ten years. Probably because it was the last, John Day was evaluated more thoroughly than earlier dams from the standpoint of migrating fish and migrating waterfowl. The photographs in this section are the most complete coverage of the stages of construction of a dam.

132. John Day Dam. 1961. Relocation work on the north shore upstream from the dam; relocation is a major part of the construction of a big dam like John Day. The work involved rails, roads, and communities, extending 80 miles. In the distance you can see two small dikes built out from the shore, the start of construction of the left abutment. US Army Corps of Engineers.

133. John Day Dam. A downstream view of the early north shore construction. Against the backdrop of the Columbia River Valley, the dam does not seem very imposing. This is the kind of scene Western artists painted by the hundreds, only without the dams, of course. US Army Corps of Engineers.

134. John Day Dam. Construction within the north shore cofferdam and excavation for the navigation lock. US Army Corps of Engineers.

135. John Day Dam. Arabesque pattern of access roads on the north shore construction site. US Army Corps of Engineers.

136. John Day Dam. North shore spillway piers, fish ladder, and lock. US Army Corps of Engineers.

137. John Day Dam. This upstream aerial view shows early construction on the north shore again, set in the spare, basalt terrain of the John Day reach of the Columbia, with towering Mount Hood in the Cascade Mountain Range in the distance. US Army Corps of Engineers.

138. John Day Dam. Closeup aerial view of the partially built spillway bays and cofferdam. US Army Corps of Engineers

139. John Day Dam. Steel reinforcements. Albert L. McCormmach.

140. John Day Dam. More steel. Albert L. McCormmach.

141. John Day Dam. Preparation for construction on the south shore. US Army Corps of Engineers.

142. John Day Dam. North and south shore five years into the construction of the dam. US Army Corps of Engineers.

143. John Day Dam. 1961. Upstream face of spillway monoliths, showing tainter gate assembly. Albert L. McCormmach.

144. John Day Dam. Construction workers passing beneath installed tainter gates. These rotary gates are sometimes called "radial" gates, but most people call them after the engineer who invented them. Ingenious and yet simple, tainter gates are cylindrically curved on the upstream side, with spokes that come down to a trunnion, or hub. Because water acts perpendicularly on a submerged surface, all of the load is transferred to the spokes, the reason the gates do not need cranes. A modest electric motor rotates them to allow water to pass under them. This water then helps in opening them further, and they close under their own weight. They are fifty feet high and fifty feet wide, and weigh 200 tons. Albert L. McCormmach.

145. John Day Dam. Gates, piers, reflections. Albert L. McCormmach.

146. John Day Dam. 1964. Downstream view of partially completed spillway bays, passing water. Albert L. McCormmach.

147. John Day Dam. 1964. Closeup upstream view of water flowing around the base of piers of the partially completed spillway bays. Piers are given careful consideration in the hydraulic design of a spillway. The spillway crest needs to be of a certain length to pass a design flood, and the piers reduce its effective length, the clear spans between piers. The crest has to be extended by the amount of the reduction, which is measured by a pier contraction coefficient. Model studies are used to determine contraction coefficients for piers of different thicknesses and nose shapes. It became standard practice on Corps of Engineer dams to make pier nose shapes elliptical, as seen here at the top. Albert L. McCormmach.

148. John Day Dam. Wheel and cable for moving a gate. As well as up-to-date electrical technology, dams make use of ancient mechanical principles. Albert L. McCormmach.

149. John Day Dam Electric Generators. The installed capacity is 2,160,000 kilowatts. Paul and Marilyn Peck.

PERSONAL AFTERWORD: WATER AND POWER

For building its dams, the Army Corps of Engineers had at its disposal the considered judgment of a large number of experienced people. I fully appreciate that now, but when I was young, I regarded those dams as pure, intense emotion cast in concrete.

Mac's mother told me that even as a boy he was always angry, no one knew why. His family thought that it was just the way he was, that some demon held him in its grip. He was not in the least religious, but if on a practical level there was a higher power for him, this deity was the power of nature, and the power of the Columbia River belonged to its court. His work on hydraulic design was his only apparent relief from the demon's grip.

Stormy moods were the usual condition of Mac, as I remember my childhood, and my usual mood was apprehension. I believed that there must be a connection between the discord at home and the river giants Mac was locked into battle with. I thought of myself as a little salmon helplessly swept down the river to be battered by rapids and shredded in the dreadful turbines. Later, after I had acquired a little knowledge, I saw myself no longer as an innocent fish tossed about, but part of the rapids themselves, responsible in hidden ways for the troubled waters that passed for normal life at home. Later yet, I recognized these imaginings as unremarkable creations of the self-absorbed world of childhood. The importance of moving water in them was their only claim to originality. The attraction to water, I now believe, came with the family.

Within a year of my birth, the Corps of Engineers first laid before the nation its plan for the Columbia Basin, its 308 Report. In some ways, the person I am is a consequence of that plan no less than are the irrigated semi-desert, the flood-protected valleys, and the power-hungry industries of the vast territory drained by the Columbia River.

For most of my life I have lived in one corner of America – in small towns in its dry interior, Pendleton, Walla Walla, Pullman, and in large, rainy cities near its coast, Portland, Seattle, Eugene. Diverse as they are in their physical surroundings, climates, and ways of life, I think of them as parts of a single region, the Pacific Northwest. The reason I can do this is the major river that flows through it, the Columbia, and the power delivered by it and its tributaries. The physical extent of the Bonneville Power Administration's long-distance power lines is a credible definition of the region, one reason why I am a Northwesterner.

The river has changed physically over time, of course, but for me the change is not just physical. First the river was the route between Pendleton and Portland, between my grandmother's house where I was happy and my own where I was not. The river was freedom. Then it became my father's work, which was secret and terrible. Then later, while I was going to college, the river became something else again. One spring I stopped by the Columbia River when it was running high from snow melt. It looked to be on a rampage. I saw it as a kind of shapeless monster, each part of which was indistinguishable from every other part, each part keeping step with its neighbors, mindlessly, predictably, sweeping up everything in its path. This was the 1950s, a time of persecution of dissent in this country, and I had just made my first conscious metaphor. The river was now powerful conformity to me. A few years later, when I worked in the hydrology section of the Corps of Engineers, just down the hall from Mac, my problem was the hydrology of Jackson Hole, the high valley formed by the Grand Tetons, which the Snake River passes through. The water that came down from there varied a lot from year to year. This work impressed me by how unpredictable a great river can be. Unpredictability being a version of freedom, it seemed I had returned to my boyhood association with the river, only I was no longer a boy, and I knew the association was deceitful. Individuals can be found who rejoice in a river allowed to run wild, but society wants a river to be useful. I was after all working for the Corps, which at the time was all about river control.

I borrow a metaphor from a poem (by Goethe), which likens the human spirit to water and human fate to wind. Water begins its course through the world with a furious, foaming waterfall, slinks across a valley floor to a quiet lake, where the wind sends it billowing and foaming again, ever changing. The beauty of the poem lies in its universality, though when I apply it to Mac I change leisurely currents to somewhat agitated ones. His career in hydraulics, the branch of science that deals with applications of moving water, a career which outwardly appears to be the outcome of historical forces and chance, the winds, can also look like a universal destiny, which might be called the human spirit. Poetry and history, each has its own insights. My image of Mac is inseparable from the course that water takes, which is both inexorable and often tempestuous.

It is a curiosity of my family that the Columbia River has had a personal meaning as a destination. Mac's father had his picture taken while standing next to Multnomah Falls, in the Columbia River Gorge, and another while standing on the bridge part way up the falls. No doubt Mac's mother took those photographs, and his father took one of her from the highway, near the foot of the falls. This may be where his father got the idea of building a house underneath or beside a waterfall. He talked about it at home, and Mac believed that they were really going to do it, live underneath a waterfall. Mac the

150. Multnomah Falls. Photograph of Mac's father, J. W. McCormmach, who was fascinated by falling water. Taken by L. B. McCormmach, Mac's mother. J. W. McCormmach.

small boy would have liked it, but he was pretty sure his sisters and mother would not have liked living around such a noisy thing. His father probably knew that too, for he also talked about moving the family onto a houseboat, a quieter way of living close to water. His father died before any move was decided. Mac wondered about his father's fascination with waterfalls, given the course his own life had taken, his work with water. A river falls a little bit all the

151. Multnomah Falls. Photograph of Mac's mother. J. W. McCormmach.

152. Multnomah Falls. Photograph of the author, 1938. Albert L. McCormmach.

time everywhere, but at a waterfall or a dam, the river falls a long way all at once; this is where Mac came in. Mac took a photograph of me at age five posed where his mother had stood, on the old Columbia Highway in front of Multnomah Falls.

No one in my family has actually lived on the Columbia River, but in waking dreams, my parents and I have. When my mother retired, she hoped to build a small house on the windy, rugged cliffs on the Washington side of the Columbia River, across from Biggs in Sherman County, near where she was born. She showed me the plan she drew for the house. She had obviously given it a good deal of thought. My father wanted to build a little house near Biggs on the Oregon side of the river, high on a rock shelf there. I do not think that my long-separated parents knew of the other's thoughts on the subject, but I could be wrong. If my parents had built their houses, they would have faced each other again, separated by the Columbia River, a physical barrier as daunting as the invisible one that had divided them when they lived in the same house. I think of a home on that river too, likewise a flight and a return.

My mother grew up on a farm in central Oregon, where there was no electricity, and no telephone. Just as the farm was about to be foreclosed in the agricultural depression of the 1920s, electric lines began to appear in the area and her family wanted to hook up, but they could not afford it. For power, they had a few hired hands during harvest, a few horses and mules, and a Fairbanks Morse gasoline engine, which did double duty by supplementing power from the windmill at the well and by running the washing machine; the water for washing was heated by coal. On laundry days her father moved the engine from the well into the house on temporary loan to her mother. The routine on her farm was a living history of power before electric power. When she got older, she took up religion, and electricity had a place in it, still a longing but for something larger. To the end of her life, she was a devoted member of a transcendental scientific church which considered electricity as immaterial and power and spirit as one and the same. Her ideas on electricity and power naturally collided with my father's, contributing in no small part to the mutual incomprehensibility at home. My father was an atheist and a materialist, for whom power was a gift of nature – in his work, it was the gift of moving water – and electric power was one of its forms; electricity was charged particles, that was all. His work had the goal of making those charged particles do useful work for us. That was what power lines were all about.

When I grew up in town, there was always electric power. At first it was just there, it came with the rest of the world. Then when I was old enough to wonder where things come from, it was mysterious and very dangerous I was told. When I began to learn about things, I came to think about it in my father's way, as a gift of nature. I know that an electric bill arrives at the end of the month, and if it is not paid, the power can be shut off, but I know it is still there waiting for me, a gift as ever. Although I now know better, I still take electric power for granted.

This book tells how through a great public works project power and water were brought together to transform the life of a region. On that same drive along the old Columbia Highway where my father photographed me in front of the falls, he took a photograph of me and my mother in front of Bonneville Dam and its reservoir on the Columbia River. It was 1938, the year the dam was finished and had just begun to deliver power, and he had just changed his kind of work and where his family lived because of that dam and because of that power. He put me in the picture of his life at that time. I am still in that picture, bound, it would seem, one day to gather his photographs of dams and water and make them into a book.

153. Bonneville Dam, 1938. Upstream view of the dam the year it went on line. The boy is the author, with his mother. Albert L. McCormmach.

APPENDIX A: HOW A POWERHOUSE WORKS

When we visit a dam, we see water drop into the intake ducts of the powerhouse, where the conversion takes place. We see it disappear into the dam, but we do not see the electricity that comes out. What we see are wires strung from the top of the powerhouse to the shore. If we do not understand the physics of it, we feel short-changed, and we definitely get no sense of the power of moving water. We do get a sense of it from river rafting, where the action is mechanical, muscular. We get a more vivid sense of it from a flood. Floods are rather common and in many places a normal occurrence. Big floods are brought on by singular events such as bizarre weather, earthquakes, and volcanoes; flash floods are particularly destructive. When people are caught in floods and cannot escape, usually they do not drown, they do not look like they simply ran out of air; rather, their skulls are crushed and their clothes are stripped from their bodies. When we have experienced a big flood or seen its destruction, we understand something important about moving water, its power. The hydroelectric power generated at a big dam is nothing so tremendous as the power of a big flood, but it is still very large.

We can get a rough idea of just how large. The big numbers below are meant to explain, not to impress. The familiar unit used to measure the power delivered by a hydroelectric dam is "kilowatt," 1,000 watts. To convert it to a more intuitive unit of power familiar to motorists, a kilowatt equals one and one third horsepower. A bright, incandescent light bulb is rated at 100 watts, one tenth of a kilowatt. Bonneville Dam, the first federal dam on the Columbia River, today has a capacity of a little over a million kilowatts; this amount of power could burn ten million light bulbs, equivalent, say, to powering a half million homes, a moderate-sized city. The corresponding unit of energy is the same as that which appears on your electric bill, "kilowatt-hour," the energy expended by a kilowatt of power in an hour's time. Another word for energy is "work." We say correctly that with a hydroelectric dam we make a river work for us.

That is because the energy to light ten million incandescent bulbs comes from the work performed by water falling inside a dam. At a hydroelectric dam, water falls by gravity through a submerged intake in the upstream face of the dam into a sloping channel called the "penstock." This channel traverses the body of the dam and enters the powerhouse below it. Water builds up pressure as it drops through it, and at the entrance to the turbines in the powerhouse, it passes through a system of louvers, known as "wicket gates." By opening and closing the gates, which are done automatically, the quantity of flow can be controlled, keeping the blades of the turbines turning at a con-

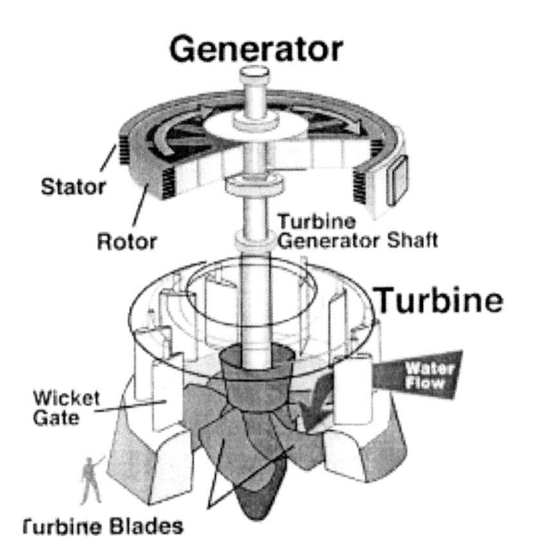

154. Electric Generator and Turbine, Drawing. Note the tiny person at the lower left, true size. US Army Corps of Engineers.

stant rate. This is necessary because the rate of turning determines the frequency of the electric current. Electric generators sit on top of the turbines. Inside of each generator, the shaft of the spinning turbine blades turns a "rotor," a series of copper wires wrapped around iron poles. Electric current is passed through the wires, converting the rotor into an electromagnet. By well-known laws of physics, the moving magnetic field of the turning rotor induces an electric current in a surrounding ring of fixed copper wires, called the "stator." This current is what the hydroelectric dam produces. Finally, being of no further use, the water that has spun the turbines is directed into the outlet tube, which has a large diameter, slowing the water before it enters the river below the dam. In summary: at a hydroelectric dam, turbines in the powerhouse convert the power of falling water into the mechanical power of rotating shafts, and the generators convert this mechanical power into electric power, which is carried by wires from the powerhouse to power lines on land.

The quantity of electric power that any given dam is capable of producing is easy to visualize. It is determined by the volume of water that flows through the turbines and by the distance the water falls, which is the distance between the level of the pool behind the dam and the level of the turbines, called the "head." The bigger the flow and the bigger the head, the greater is the power produced; this is both intuitively credible and scientifically exact. The powerhouse containing the turbines is located as low as possible to maximize the head. Only the top of the powerhouses at Columbia and Snake River dams can be seen above tailwater. Each powerhouse contains a number of generat-

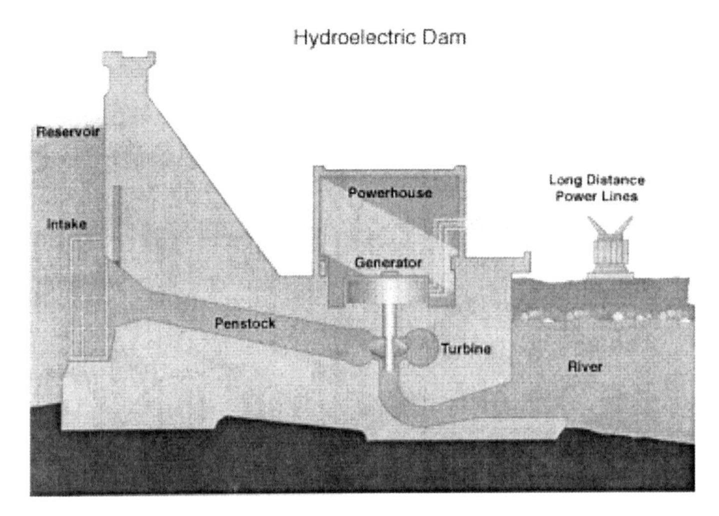

155. Hydroelectric Dam, Cross Section. This diagram shows how the power of water falling from the reservoir through the dam and powerhouse is converted to electrical power carried by power lines. Tennessee Valley Authority.

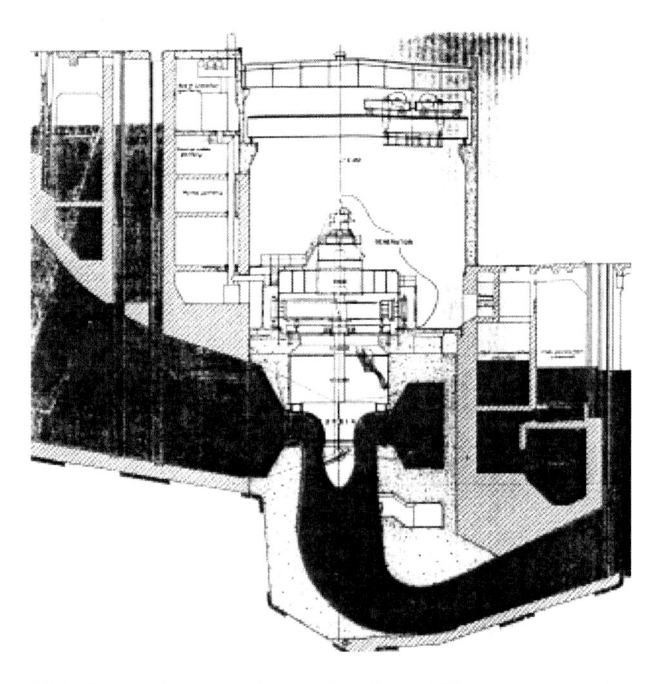

156. Powerhouse, Cross Section. Water enters the powerhouse through the penstock on the left, falls through the turbine, and exits downstream to the right. The galleries to the left of the generator are for electrical wiring, control cables, and pumps. The upper submerged channels on the right are for collection and transport of upstream migrating salmon, intended to prevent them from trying to enter the powerhouse. The Dalles Dam. US Army Corps of Engineers.

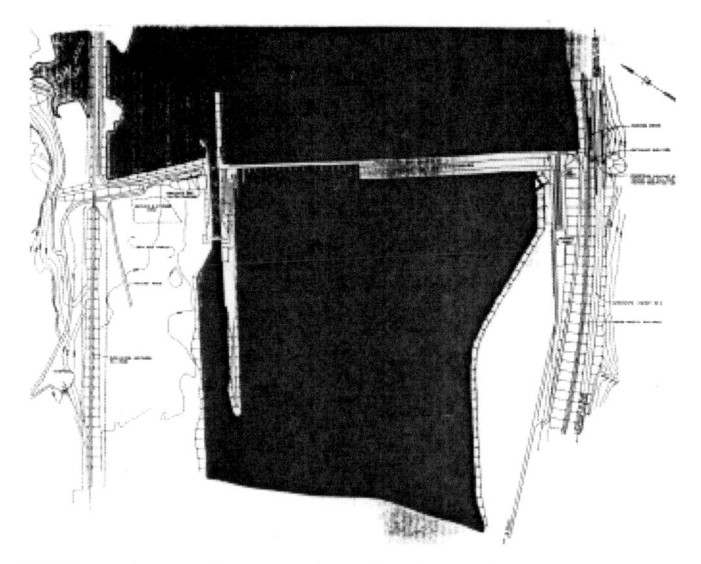

157. Powerhouse, Dam, and Lock, Plan. The powerhouse is to the right, extending half way across the river. The spillway and lock, the other half, are to the left. Railroads run along both banks and the interstate along the right bank. John Day Dam. US Army Corps of Engineers.

ing units, and the number can change. Bonneville Dam is typical: its first two units went on line the first year, and four others did so at intervals over the next four years. The more generating units there are, the more water that falls through them, and the greater the power produced. The number of powerhouses at any given dam can change, too, and with it the total power produced. We have seen that Bonneville added a second powerhouse and Grand Coulee a third powerhouse to its original two. With its three powerhouses and full complement of generating units, Grand Coulee Dam could light sixty-five million incandescent bulbs or power a metropolis.

APPENDIX B: ON POWER LINES

Inside the powerhouse of a dam, as we have seen, water turbines and electric generators convert the energy of falling water into electric energy, which is conveyed to consumers over power lines. The lines usually have to be long, since a powerhouse is in a fixed location on a river, often remote, and consumers are widely distributed. America's 157,000 miles (as of 2002) of high-voltage (upward of 230,000 volts) long-distance power lines are a conspicuous presence over vast tracts of the country. There are in addition hundreds of thousands of miles of low-voltage, regional and local power lines.

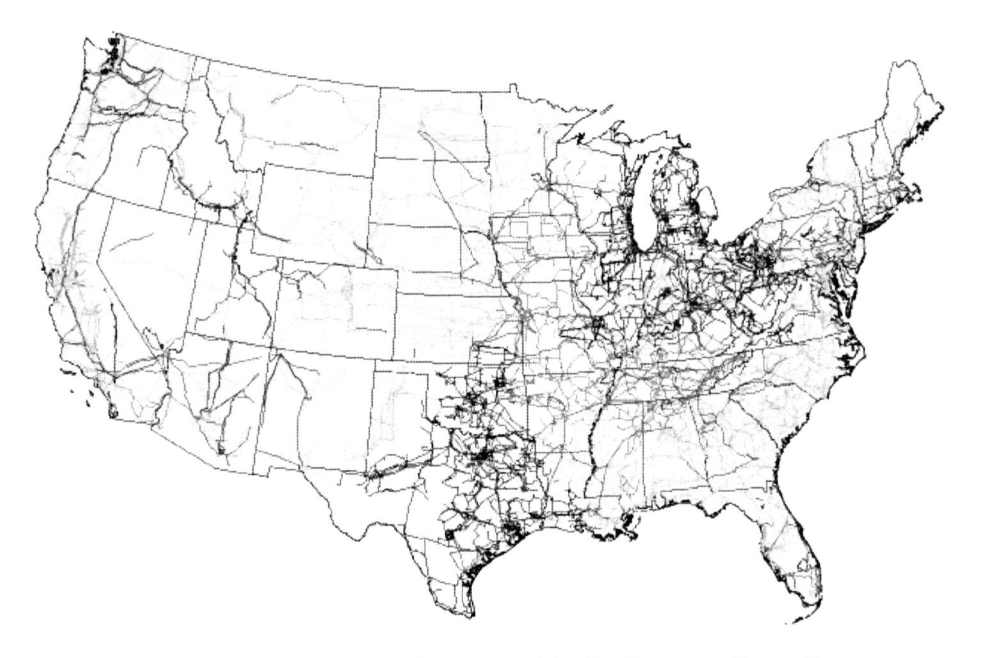

158. US Transmission Grid, Map. The geographic distribution of long-distance transmission lines in the US is highly uneven, as this map shows. Line voltages vary from 115,000 volts (faintest lines on the map) to 1,00,000 volts. Further on in this appendix, there is a discussion of "grid's." There is no national electrical power grid; rather the US is served by three, largely separate major grids, or "interconnections," mostly owned by local utilities: Eastern, Western, and Texas. The Eastern and Western interconnections extend into Canada; likewise, a Quebec interconnection extends into the US. Federal Emergency Management Agency.

Electric power can be transmitted underground and undersea or overhead. Transmission over long distances on land makes use of overhead lines, which are cheaper to construct than underground lines, and air is a better insulator than earth. Overhead lines are generally used for short distances as well, though here the use of underground lines is growing; overhead lines take up space, and they are regarded as an intrusive feature of a landscape.

Low-voltage, overhead power lines are strung from insulators attached to the side of poles or placed on top of cross arms of poles. The poles, which in the US are thirty to forty feet high, are usually made of wood but they can also be made of concrete, metal, or reinforced plastic. High-voltage transmission lines over long distances are supported by strings of porcelain, glass, or composite polymer insulating discs – the number of discs depends on the voltage on the lines – suspended from cross arms of towers, or pylons. The number of pairs of cross arms and the particular shape of the towers vary according to the number and type of circuits they carry and other design criteria, but everybody recognizes towers for what they are; there is nothing like them. Most of the towers we see today were built in the 1950s and 60s when the standard design was a steel lattice; since the 1990s, other designs have become common. Because persons and other objects must not come near the conducting wires (and also for access), the area under them is cleared as a right-of-way to a width of seventy-five to 150 feet or even more. For the same reason the wires have to be raised high off the ground. Towers commonly stand about as tall as a ten-storey building, though depending on terrain they can rise very much higher, the record tower at a Yangtze River crossing measuring over 1,000 feet. High-voltage lines normally contain two to six wires in parallel, separated by insulating spacers; the wires themselves, called "conductors," are bare and heavy, made of an aluminum alloy twisted around a steel core for strength. Copper is a better conductor and is sometimes used for low-voltage transmission and grounding, but aluminum is lighter and cheaper, for which reasons it is the conductor of choice for high-voltage, long-distance lines.

The original Edison electric installation used direct current, which had serious drawbacks. Direct current was produced and delivered at fixed voltages, and there was no common means of varying the voltages; this meant that different uses of direct current – lights, motors, and streetcars – employing different voltages required different generators and circuits, an inconvenience and an expense. The second drawback of direct current had to do with an excessive loss of electric energy to heat, a consequence of the low voltages that were used, again an expense; direct-current voltages normally were kept low because of the difficulties and dangers of delivering high voltages to small loads. The only way heat losses in Edison direct-current transmission could be reduced was to use heavier wire, which was made of copper, another expense. To get around these drawbacks, two things had to happen. The first was to replace direct current with alternating current, or current which changes direction in a regular manner, the measurement of which is frequency, which in the United States is sixty cycles per second. The second was to introduce the transformer, an ingenious device consisting of two windings of insulated wire and a magnetic core, through which alternating current is passed. The function of the transformer, which physics explains by the principle of electromagnetic induction, is to vary the current and voltage of an alternating current without changing the power and the frequency.

Transformers and alternating currents enable the same circuit to be put to different uses. With transformers to supply alternating current at the voltages required by the customers, a single generating plant can meet every need, a savings. The first drawback of direct current is removed.

For motors as well as lights to run off the same circuit, another early development was helpful, the three-phase circuit. In a three-phase circuit, three wires carry current of the same frequency with different phases; that is, the oscillations of current in the three wires reach peak values at different times. The spacing of peaks is exact: if we take one current as the reference current, the other two currents are delayed by one third and two thirds of an oscillating cycle. The delays make possible the production of rotating magnetic fields, which simplify the design of electric motors, a principal use of alternating currents. The three-phase circuit has additional advantages over circuits carrying only one or two phase wires: it provides a steady delivery of power over a cycle with minimum energy loss, and it needs less wire to transmit power. It is the type of circuit commonly used for electric power transmission.

Transformers to vary the voltage on a line are the key to removing the

159. Three-phase Electric Power Transmission. We see three rows of towers sharing a common right-of-way. They are called "delta" towers because of their shape, designed to carry the cross arm at the top. The lines here are bundled, two conductors for each of the three phases. Towers with a single cross arm like these carry a single circuit; towers with three cross arms carry two circuits, three lines on each side. Both types are common in the Northwest. Wing-Chi Poon.

second drawback of direct current, the unacceptable loss of electric energy to heat accompanying low-voltage transmission. The explanation rests on two well-known physical laws: the heat generated in a conducting wire is proportional to the square of the current, and the power delivered over the wire is proportional to the product of the voltage and the current. It follows that if for a given power the voltage is increased, the current is decreased in the same proportion, and the heat loss is decreased in greater proportion. A numerical example makes this clear: if the voltage is doubled, the current is halved, and the heat loss is quartered, the transmitted power remaining unchanged. The loss of electric energy to heat is reduced to acceptable levels by transmitting electricity at the high voltages made possible by transformers.

The longer the power line, the more electric energy is lost to heat, and the higher the desirable voltage is. Alternating current was first transmitted in 1886 over a short distance, using a low voltage, 1,000 volts. Five years later, it was transmitted over a longer distance, 110 miles, using a medium voltage, 25,000 volts. Power lines lengthened, and voltages increased, reaching 150,000 volts by the First World War, and 230,000 by 1923. That, as we have seen, was the voltage on the lines from Bonneville and Grand Coulee Dams to the cities on the Pacific coast in the 1930s, by which time even higher voltages had been used elsewhere. For lines connecting powerhouses, 1,000 volts per mile has been a rule of thumb, but much higher voltages are used today on long-distance lines. The Bonneville Power Administration (BPA), having made many studies of high-voltage alternating-current transmission, increased the standard voltage on its new regional power lines to 500,000 volts in the 1960s. The reason was the same: at the higher voltage, electric energy is transmitted with a smaller current, which results in a smaller loss of energy due to the heating of the conducting wire and also in less damage to equipment caused by the heating. Today, around the world, voltages on long-distance lines vary from around 100,000 volts to over 1,000,000 volts, most falling in the middle range, the choice being largely an economic decision.

High-voltage transmission works this way. Moderate voltages at powerhouses and other generating plants are stepped up by transformers for electric transmission over long-distance power lines. At the receiving end, the voltages are stepped down by other transformers for distribution over local lines to consumers. In the beginning, only one transformer was used to step down the voltage, but with the arrival of very high transmission voltages, it was found convenient to use two or more. Nowadays a long-distance line terminates in a substation – a bank of transformers lowering the voltage to an intermediate value – where distribution circuits serving the region tap into heavy conductors, or bus bars. Voltage is again lowered by smaller substations or local cylindrical transformers mounted on utility poles at the locations of customers.

The electricity we plug into to run our houses is "split-phase," providing 120 volts for most uses and 240 volts for large appliances. In summary: with the help of transformers at each step, electricity from major power producers is transmitted over long distances to their terminus, where it branches into distribution lines, from which it branches again into feeder lines entering homes and businesses. This is how most electricity is moved today, making use of centralized electric power generation and transmission. Power, if small in scale by comparison, is also generated at the local distribution level, for example, by rooftop solar panels and wind turbines, which can be brought on line.

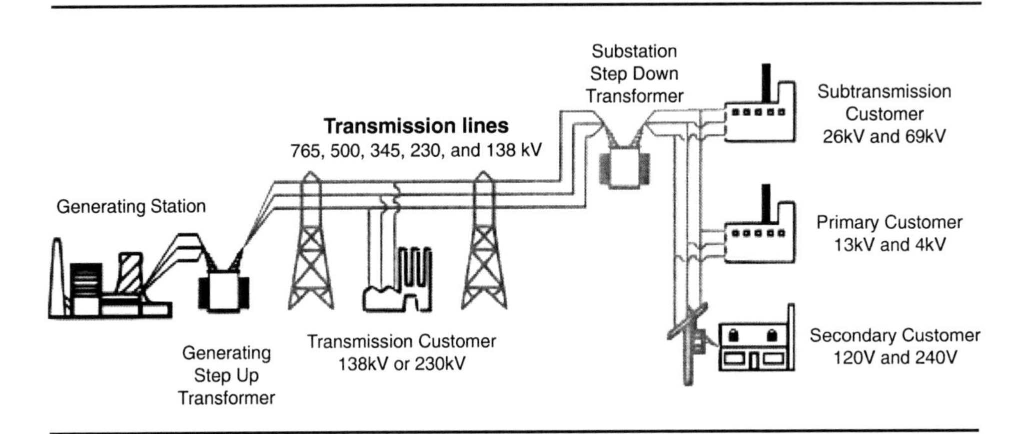

160. Long-distance Power Line, Drawing. US Department of Energy. The three lines represent three-phase power.

An electric "grid" is a network for delivering electric power from suppliers to consumers. The term is flexible, its meaning varying with context: we speak, for example, of a "national grid" meaning the overall network, a "regional transmission grid" meaning a network of long-distance lines, a "local utility distribution grid," and so on. Circuits connecting hydroelectric powerhouses with one another and with other electrical generating plants powered by coal, nuclear fission, gas, oil, or alternative sources of energy form generation grids. Comprehensive grids include power generators, transmission lines, substations, and distribution lines. Delivery of electric power using grids has a number of advantages over delivery by individual circuits; for one, because consumers do not all use electricity at the same time, the overall generating capacity can be smaller than it would be if every consumer has its own circuit. Other advantages are greater reliability of service and efficiency. With a grid arrangement, electric power is routed to users over a variety of redundant paths, balancing the load and allowing for line failures. Because long-distance transmission of electric power is cheap, it is often more economic to draw power from a dis-

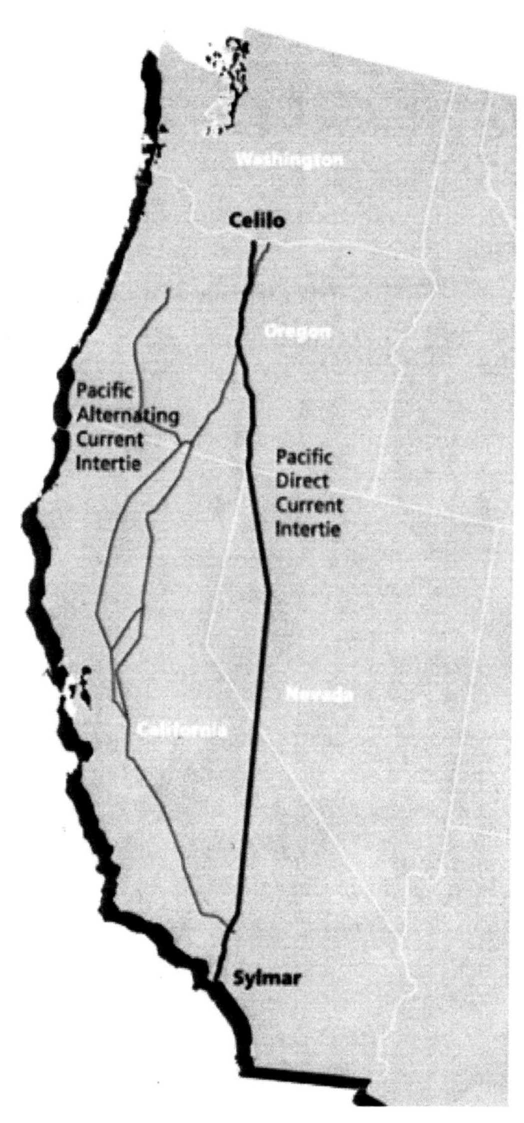

161. Pacific Intertie, Map.
Bonneville Power Administration.

tant source than from a local one.

During the Kennedy Administration, the interior department assigned a task force headed by the BPA administrator to study a possible interconnection of transmission grids in the Pacific Northwest and Southwest. In keeping with its recommendations, in 1964 Congress appropriated initial funding for the Pacific Intertie, destined to coordinate federal, public, and private electric power systems in eleven Western states and British Columbia, the largest-ever single transmission program in America, a super-grid. The project was carried out by the BPA, the Bureau of Reclamation, the city of Los Angeles, and several utilities. The heart of the intertie is four very long, very high voltage transmission lines, three of which are alternating-current lines and one is a direct-current line. Completed in 1968 and 1969, two 500,000-volt alternating-current lines proceed from a station on the Lower Columbia River at John Day Dam to a station near Los Angeles, a distance of around 900 miles; a third 500,000-volt alternating-current line connecting the Northwest power grid to a station near San Francisco was completed in 1993. The direct-current line proceeds from a station at Celilo near The Dalles Dam to a station at Sylmar near Los Angeles; at each station, alternating current is converted to direct current, which is sent to the other station, where it is reconverted to alternating current for distribution. In 1985, the voltage on the direct-current line was doubled to 800,000, and today is 1,000,000. The intertie transmits prodigious quantities of electric energy, up to 4,800 megawatts on its alternating-current lines and 3,100 on its direct-current line.

The direct-current Pacific Intertie line became practical with the development of large mercury-arc rectifiers, which are vacuum tubes for converting alternating to direct current, with computers synchronizing the conversion at both ends. At high-voltages, direct-current transmission has advantages over alternating-current: lower energy losses, less environmental interference, and lower cost, since it takes only two wires instead of three. The BPA was instrumental in this major breakthrough, having begun research on direct-current, long-distance transmission two years before the intertie was approved.

It is generally recognized that America's electric grids are inadequate. Demand for electric power has grown by twenty-five percent since 1990, while investment in new high-voltage transmission lines has fallen by thirty percent. The result has been grid congestion, costly energy loss, and obsolescence in an increasingly competitive electric power market. As well, low voltage distribution has proven unequal to rapidly changing consumer needs. When the present grids were installed, our society was at a different technological and economic stage than it is today.

Demand for and production of electricity have to balance at all times, or outages will occur or there will have to be means for storing electric power. Our existing electric grids meet changes in power demand by manually ramping up or down our standard electric generating plants, which can be relied on to deliver, having a known, steady capacity. Power delivered by alternative sources such as wind and sun, because of their intrinsic unpredictability, does not fit this scheme. Owing in part to our increasing interest in these sources, there is a movement today to replace the manually operated electric grids with computer-operated ones, so-called "smart grids."

Obama as incoming president supported the changeover to smart grids from the start, asking Congress to pass legislation to upgrade the nation's long-distance transmission and local-distribution grids making use of digital technology. His economic stimulus bill directed $10.5 billion to this end, a down payment on a long-term project. A popular analogy is with Eisenhower's initiative in upgrading America's road system with an interstate freeway system, fifty years in the making. Just as interstate highways were important for the manufacturing society of the 1950s, the analogy goes, a modern electric grid is important for today's information economy. The cost is comparable too, with estimates varying between $100 billion and $2 trillion, though unlike the highways, which the federal government paid for, the grid would be financed largely by the private sector. Principal goals of a smart grid are to optimize the use of electricity and to stimulate alternative-energy production. Other benefits are conservation of electric energy, lower consumer costs, greater reliability of service, lower carbon-dioxide emissions, resistance to enemy attack, and creation of jobs in the green-energy industry. Smart grids evolved from innovative methods in metering, monitoring, synchronizing, and automatically controlling loads, and again from research at the BPA, which revolutionized these methods as they apply to very large geographical regions. Europe, China, Japan, Canada, and other countries are working on smart grids, which clearly hold the future for modern societies with their strong dependence on electricity.

The wish that Woody Guthrie made in his song celebrating the birth of Columbia Basin power, this book's epigraph, "What I think is the whole world oughta be run by E-lectricity," is, as far as modern societies go, approaching fulfillment. At the time Guthrie wrote the song, when Bonneville Dam was just starting out, only ten percent of America's energy was used to generate electricity. In 1970, at the close of the era of building big dams in America, the figure was twenty-five percent, and today it is forty percent. The unique ability of electricity to convey both power and information promises that the percentage will continue to rise. America's prosperity, security, and health rely on electric power, and this means more and better electric power grids.

At the time of writing, this country is recovering from a severe recession, likened to the Great Depression. We have seen that Roosevelt approved and funded federal power development in the Pacific Northwest to provide jobs for the unemployed, among other reasons. The first long-term returns were Bonneville and Grand Coulee Dams and an agency to build the transmission grid and market the power, the BPA. Early in 2009, Obama signed the American Recovery and Reinvestment Act, under which the BPA, with $3.25 billion additional funding to draw on, embarked on a new transmission project, sev eral years in the planning, a 500,000-volt transmission line between the McNary Dam station and the John Day Dam station. The agency has plans for three more like projects, for a total of 225 miles, to be added to its existing 15,000 miles of transmission lines (three quarters of the Pacific Northwest's high-voltage grid). The object is to create jobs, to enable the Northwest transmission grid to meet increased energy demands, and to incorporate new sources of renewable energy such as wind, into which the BPA invests and does research. In this respect, this book ends where it begins, with a recovery act promoting federal power development in the Columbia River Basin.

APPENDIX C: IMAGES OF POWER

Because this is an illustrated book about power, a word about images of power is in order. Their intention and effect vary greatly. Civil rights marches in the American South give us images of the power of a just cause and moral persuasion. At the other extreme, owing to a common association of power with brutality, force, and wealth, images of power can seem odious to us. Nothing excited Hitler – who trusted only one motive, power, everything else being bosh – so much as contemplating designs and models of gigantic, triumphal buildings, made to order by his pet architect, to be erected in Berlin following the Third Reich's victory in World War II. These images convey domination and a leader's absolute power. Images of dams, electric power generators, and power lines can suggest domination too, in this case domination over nature, in this time of environmental awareness, but these same images can seem seductive during an economic depression and compelling during a war for national survival. They can also inspire artistic responses in certain people. The subject is touched on earlier in this book in connection with the architecture of dams. Here the discussion is extended to the fine arts.

With his active interest in photography and his technical employment, Mac naturally knew something about the work of the contemporary great American industrial photographer and painter Charles Sheeler, 1883-1965. More than likely Mac first came across Sheeler in *Fortune* magazine, which had commissioned Sheeler to paint six scenes celebrating industrial power to illustrate a series of articles on power in America. The subject was topical. Sheeler received his commission in 1938, the year that power from Bonneville Dam was started up, which was also the year Mac went to work on power lines for the Bonneville Project. The paintings were brought together as a portfolio of reproductions, "Power," published as a supplement to the December 1940 issue of the magazine. The accompanying text says that in these paintings, machines appear not as "strange, inhuman masses of material, but exquisite manifestations of human reason." Machines are to modern artists like Sheeler what the human figure had been to artists of the Renaissance, the text goes on to say; Sheeler likened locomotives to Chartres Cathedral, of which he made a series of photographs. To persons today who despair of industrial pollution, any artistic appreciation of technology, "machine art," may seem like denial, like putting lipstick on a pig, but to most Americans in the 1930s, environmental concerns were not yet pressing, and machines were a hopeful product of human reason and appealing objects to behold. (They still are to many of us.)

Acclaimed for his series of industrial photographs of Henry Ford's River Rouge plant outside Detroit ten years before, Sheeler was the perfect candidate for the *Fortune* series. To prepare, he spent a year traveling, visiting power plants and keeping a photographic record, which he used in place of sketches. Photographs provided him with an accurate starting point, but his paintings are not exact copies. They are abstract idealizations of the manmade world, emphasizing the geometric forms rather than the dynamism of machinery. Their values are artistic; the beauty of machinery being intrinsic, a product of its function, they make no social commentary, or at least the painter intended none. Sheeler was drawn to Cézanne and the Cubists, whose formal innovations can be seen in his work. Physical technology offered him ample subjects for painting, and photography, an optical technology, offered him his signature technique in painting. In the history of art, Sheeler is known as a Precisionist.

Sheeler's six images of power are: "Primitive Power," an early waterwheel; "Rolling Power," the running gear of locomotives; "Yankee Clipper," an airplane propeller; "Steam Turbine," a fossil-fuel power generator; "Conversation between Sky and Earth," power lines, insulators, and steel tower depicted against the background of Hoover/Boulder Dam; and "Suspended Power," a water turbine being lowered into place in the powerhouse of a Tennessee Valley Authority dam. The last two of the series are reproduced below. They represent a different perspective on the industrial giantism of the subject of this book, hydroelectric dams and networks of power lines.

162. "Conversation between Sky and Earth." Oil painting by Charles Sheeler, 1939. Power lines and a dam, a superposition, both a reality and a formal pattern. The title and the perspective draw the eye to the soaring elevation of Hoover/Boulder Dam and of the power-line tower that goes with it. The dam is painted in a warm tan, close to the color of flesh, a suggestion perhaps of the humanity of technology. Curtis Galleries, Minneapolis.

163. "Suspended Power." Oil painting by Charles Sheeler, 1939. A water turbine that will drive an electric generator is suspended above the floor of the powerhouse, ready to be lowered into place. The location is Guntersville Dam, a hydroelectric dam on the Tennessee River in Alabama, owned and operated by the Tennessee Valley Authority. The generating capacity, 140,000 kilowatts, does not begin to compare with that of dams on the Columbia River, but it served Sheeler's purpose. Here he made an exception of his practice of leaving people out of his paintings – he thought that the world would be more beautiful without people in it – by including three workmen, no doubt to show that on the scale of powerhouse machinery they are precarious objects. Again the painting is done in a flesh tone. Dallas Museum of Art.

APPENDIX D. RECENT IMPROVEMENTS IN FISH RETURNS

If, in places, this book conveys a pessimistic outlook for migratory fish in the Columbia Basin, that is because it follows the course of Mac's thinking. After Mac retired, in the 1970s and 80s, the long-term prospects for salmon in the rivers looked unpromising to many close observers, and so they looked to Mac. I know this because I was there, and he talked about it. He had grown despondent over prospects of survival of salmon in rivers filled with dams. I do not know what his thoughts on the matter were in his last decade, in the 1990s when prospects began to look up.

He was hopeful when he designed fish facilities for the dams. Later, as results came in, he lost hope. I remember well his use on occasion of words such as "tremendous disaster" when talking about the hazards dams posed for fish returns over the long run. Today much more is known about the problem, and the prospects for fish and dams appear reasonably hopeful again, at the very least more encouraging. The following observations are in the way of a brief update.

Mac's expectations for fish survival in the developed rivers of the Pacific Northwest were formed at a relatively early stage of development of fish facilities and before changes were made in the operation of dams to make them less hazardous to fish runs and before the fresh-water habitat of the fish was improved. He retired a few years after the dedication of John Day Dam, in the early 1970s, around which time fifteen percent of juvenile fish were lost at each dam, while today the loss is down to five percent, and the figure for adult fish is under one percent. In recent years, Chinook salmon and sockeye salmon have returned in record numbers, and steelhead runs have improved as well. To take one example, in 2010 a record number of fall Chinook adult salmon were counted at the Corps of Engineer's last dam on the Snake River, Lower Granite Dam, and also a record number of egg nests were counted on its tributaries. It is clear that a good many of the problems of dams have been corrected. The outlook has definitely improved since Mac was engaged.

The reasons for the rebound of migratory fish stocks in the Columbia Basin since Mac's time are several, which include, for juvenile fish: recent installation of removable spillway weirs, discussed above; transfer of hatchery-born fish to river sites for their final rearing before release (the majority of returning adults are hatchery fish, it should be noted); installation of bypass systems developed over a thirty-year period, which screen seventy to eighty-five percent of the fish from turbine intakes; and spilling water at the dams to flush fingerlings downstream. Over the last three decades, the Corps of Engineers and the BPA have invested over $100 million a year in research on the problem. This has led to a better understanding of the tangled causes of fish mortality, which today's critics sometimes lump simplistically with hydropower and blame entirely on the Corp's dams. Hydropower dams are one of the hazards the fish face in the region's rivers, each of which has been addressed with varying success.

For myself, I am guardedly optimistic. In part this comes with living in the Pacific Northwest, which despite its share of economic hardships is an optimistic land. Here I am regularly reminded of the bounty of hydropower and of other benefits of dams. When I was growing up, my family's income came from the development of rivers. My house is heated with a furnace run by inexpensive hydropower, and it is always dry, as it would not be without dams to control the floods endemic here. These things might be discounted as merely personal, though I think they have some generality; for the overall prosperity I see around me owes much to the development of the rivers of the region. I should be faulted if I did not feel a gratitude. I also know the downside of the development of rivers, even if my experience of it is not so immediate. I believe that when everything is taken into account, the good and the bad, the development of the rivers of the Pacific Northwest has been beneficial in the balance. In support, I lean on my chosen field, history, which has a measure of circumspection built into it.

LIST OF ILLUSTRATIONS

BIBLIOGRAPHY

Below are several excellent books I found helpful.

Billington, David P., and Donald C. Jackson. *Big Dams of the New Deal Era: A Confluence of Engineering and Politics.* Norman: University of Oklahoma Press, 2006.

Cone, Joseph. *Common Fate: Endangered Salmon and the People of the Pacific Northwest.* Corvallis: Oregon State University Press, 1996.

De Villiers, Marq. *Water: The Fate of Our Most Precious Resource.* Boston, New York: Houghton Mifflin Company, 2001.

Goldsmith, Edward, and Nicholas Hildyard. *The Social and Environmental Effects of Large Dams.* San Francisco: Sierra Club Books, 1984.

Hildyard, Nicholas. See Edward Goldsmith.

Jackson, Donald C. See David P. Billington.

Kalez, Jay J. *Harnessed Waters: Legend of the Mighty Columbia River of the Setting Sun.* Spokane: Lawton Printing, n.d.

McCully, Patrick. *Silenced Rivers: The Ecology and Politics of Large Dams.* London: Zed Books, 1996.

McNeil, William. *The Rise of the West: A History of the Human Community.* Chicago and London: University of Chicago Press, 1963.

Merchant, Carolyn, ed. *Major Problems in American Environmental History.* Lexington, Toronto: D. C. Heath and Company, 1993.

Peterson, Keith C., and Mary E. Reed. *Controversy, Conflict and Compromise: A History of the Lower Snake River Development.* Walla Walla: Walla Walla District, US Army Corps of Engineers, 1994.

Pitzer, Paul C. *Grand Coulee: Harnessing a Dream.* Pullman: Washington State University Press, 1994.

Preston, Howard A. *A History of the Walla Walla District.* Part 1: *1948-1970.* Part 2: *1970-1975.* Walla Walla: Walla Walla District, US Army Corps of Engineers, 1970, 1975.

Rap, Valerie. *What the River Reveals: Understanding and Restoring Healthy Watersheds.* Seattle: The Mountaineers, 1997.

Reed, Mary E. See Keith C. Peterson.

Shallat, Todd. *Structures in the Stream: Water, Science, and the Rise of the U.S. Army Corps of Engineers.* Austin: University of Texas Press, 1994.

Smil, Vaclav. *Energy at the Crossroads: Global Perspectives and Uncertainties.* Cambridge, MA, and London: MIT Press, 2005.

Smith, Norman. *A History of Dams.* Secaucus, New Jersey: Citadel Press, 1971.

Springer, Vera. *Power and the Pacific Northwest: A History of the Bonneville Power Administration.* Washington, DC: US Department of the Interior, 1976.

Stacy, Susan N. *When the River Rises: Flood Control on the Boise River, 1943-1985.* Boulder: Institute of Behavioral Sciences, University of Colorado, 1993.

Tobey, Ronald C. *Technology Is Freedom: The New Deal and the Electrical Modernization of the American Home.* Berkeley, Los Angeles, and London: University of California Press, 1996.

US Department of the Interior, Bureau of Reclamation. *Dams and Control Works.* 3d edition. Washington, D.C.: US Government Printing Office, 1954.

Willmingham, William F. *Water in the "Wilderness": The History of Bonneville Lock and Dam.* Portland: Portland District, US Army Corps of Engineers, n.d.

Worster, Donald. *Rivers of Empire: Water, Aridity, and the Growth of the American West.* New York: Pantheon Books, 1985.

Wright, Richard. *The Organic Machine.* New York: Hill and Wang, 1995.

INDEX

Italicized numbers refer to pages with illustrations.